"Uniting strategy, governance, and technical foundations, this book is a rigorous guide to building and leading effective AI organizations."
Maharaj Mukherjee, Senior Vice President and Senior Architect Lead, Bank of America

"This book isn't about algorithms or the latest buzzwords. It's a field guide for leaders who want to make AI deliver lasting business impact. By grounding the conversation in strategy, adoption, and change management, *Leading Enterprise AI Programs* demonstrates how to avoid the traps that derail most projects. Every executive and AI leader should keep this book within arm's reach."
Dhivya Nagasubramanian, AI Leader and Vice President AI Transformation and Innovation, U.S. Bank

"An AI playbook for senior leaders charged with scaling AI for measurable business value. Drawing on deep operational insight, Patrick Bangert captures exactly what I've seen in Fortune 100 environments. The book is a must-read for any executive serious about embedding AI across a complex, regulated organization."
Fauzia (Foz) Saeed, VP Data Product Strategy and Governance, McKesson

"Clear and practical, this book is an essential guide for anyone launching an AI initiative, answering every pertinent question from project kickoff to full-scale deployment. Thoughtful and meticulous, it tackles every key question on AI—from technology and ethics to change management and deployment—making it an indispensable resource."
Sean Im, Founder and CEO, VerticalAI

"Tackles enterprise AI practically and realistically. It's written for leaders who are curious about where to start, managers juggling multiple AI projects, and practitioners or startups building these systems from the ground up. It openly discusses where AI projects fail and why, providing a realistic view of what you're getting into. From your first strategy meeting all the way through to maintaining operations years down the line, this book guides you through the entire journey without sugarcoating."
Temi Odesanya, Responsible AI Leader, AIG

"A rare blend of strategy, structure, and empathy that captures what it truly takes to lead AI programs that deliver impact, trust, and long-term value."
Srujana Kaddevarmuth, Global Head Applied AI Labs, Walmart

"From navigating organizational resistance and securing executive buy-in to building scalable data infrastructure and establishing governance frameworks that actually work, I have found the practical nature of this book to be invaluable. For anyone leading or contributing to data and AI initiatives at large organizations, this book offers the kind of wisdom that typically takes years of hard-won experience to accumulate. It's become an essential part of my professional toolkit, and I suspect I will continue to reference it again and again."
Kristine Swan, VP Digital Strategy and Innovation, Phillips 66

"Nails the hardest part of enterprise AI: turning strategy into sustained value. As someone leading AI in a complex enterprise, I found this book refreshingly practical. It cuts through the hype and lays out exactly what it takes to build AI programs that deliver real value at scale. It's a must-read for leaders who want AI to actually land."
Sarah Karthigan, Vice President of Artificial Intelligence, Weatherford

"Patrick Bangert is the real deal. He combines world-class AI expertise with the discipline to turn it into value. In a field full of noise, he's the signal."
Christian Anschuetz

"Patrick Bangert turns a complex topic into a practical roadmap any enterprise can follow with confidence for democratizing AI. He shows that governance is not bureaucracy but the backbone of adoption and reminds us that leading AI is not about algorithms, but about aligning people, purpose, and processes until innovation becomes inevitable. It is all about what actions get us closer to the value and for enterprise AI programs it's how to practically bring AI to data rather than data to AI."
Vinod Nair, Vice President of Data Platform Engineering and Architecture, Comcast

"A rare work that bridges strategy design and execution realities. It unites the rigor of management doctrine with the fast-evolving challenges of AI, making it an indispensable playbook for leaders navigating transformation."
Brook Selassie, Vice President, CIO and Corporate Strategy Practice, Gartner

"This book is more than a guide to leading enterprise AI—it is a blueprint for building trust, clarity, and lasting value in a rapidly evolving world. It transforms complex ideas into practical wisdom, empowering every reader to lead with purpose and confidence."
Ashwini Ghogare, GenAI Leader for Start-ups, Life Sciences and Healthcare, Amazon Web Services (AWS)

"The holistic initiative driven approach proposed in this book will increase the effectiveness of any enterprise AI program and increase its chance of delivering lasting value."
Sam Hamilton, former SVP Data and AI, Visa

"Patrick Bangert has delivered a clear and optimistic blueprint of turning vision into value, based upon teamwork, user focus, and ethical leadership. By building trust and fostering continuous learning, he shows how AI can become a powerful force for transforming the organization."
Andy Flowers, Enterprise AI Manager, ConocoPhillips

"In a world where AI hype often overshadows real impact, this book is a breath of fresh air. Patrick Bangert cuts through the noise, offering a blueprint that is as pragmatic as it is visionary. The book reminds us that sustainable AI success isn't about chasing the latest algorithm; it's about building trust, focusing relentlessly on business value, and empowering people at every level of the enterprise. Whether you're just starting your AI journey or scaling a mature program, you'll find wisdom here that's grounded in experience and delivered with clarity. I'm grateful for Bangert's candor and leadership. This is the handbook I wish I'd had years ago."
Nasim Eftekahri, Chief AI Officer, City of Hope

Leading Enterprise AI Programs

Optimize AI Teams for Value Creation

Patrick Bangert

KoganPage

First published in Great Britain and the United States in 2026 by Kogan Page Limited

Kogan Page
Kogan Page Ltd, 2nd Floor, 45 Gee Street, London EC1V 3RS, United Kingdom
Kogan Page Inc, 8 W 38th Street, Suite 902, New York, NY 10018, USA
www.koganpage.com

EU Representative (GPSR)
eucomply OÜ, Pärnu mnt 139b -14 11317, Tallinn, Estonia
www.eucompliancepartner.com

Kogan Page books are printed on paper from sustainable forests.

ISBNs
Hardback   978 1 3986 2321 7
Paperback   978 1 3986 2319 4
Ebook   978 1 3986 2320 0

British Library Cataloguing-in-Publication Data
A CIP record for this book is available from the British Library.

Library of Congress Cataloging-in-Publication Data
A CIP record for this book is available from the Library of Congress.

Typeset by Integra Software Services, Pondicherry
Printed and bound by CPI Group (UK) Ltd, Croydon CR0 4YY

CONTENTS

Online resources can be found at www.koganpage.com/leap

ABOUT THE AUTHOR

Patrick Bangert is the Chief of Artificial Intelligence at Occidental Petroleum where he leads a global cross-functional team to improve many physical and human processes by leveraging advanced analytics and AI. His thought leadership on AI has won him several awards and makes him a popular speaker at conferences and events.

Previously, he was Senior Vice-President for data, analytics, and AI at Searce, which provides professional services. He headed the profit center that is responsible for all projects with a data scientific character globally. Before Searce, Patrick was the vice-president for corporate strategy and the vice-president for AI at Samsung SDS where he led the AI Division from 2020 to 2023, bringing AI tools and services into Samsung Cloud for computer vision, natural language processing, and machine learning with a particular focus on medical imaging.

As a volunteer, Patrick spent six years co-directing the Digital Energy Technical Section of the Society for Petroleum Engineers (SPE) and published their quarterly newsletter.

Before joining Samsung, Patrick spent 15 years as CEO at algorithmica technologies, a machine learning software company serving the chemicals and oil and gas industries. Prior to that, he was Assistant Professor of Applied Mathematics at Constructor University in Germany, as well as a researcher at Los Alamos National Laboratory and NASA's Jet Propulsion Laboratory. Patrick obtained his machine learning PhD in mathematics and his Masters in theoretical physics from University College London, and his business degree from INSEAD.

A German native, Patrick grew up in Malaysia and the Philippines, and later lived in the UK, Austria, Nepal, and the US. He has done business in many countries and believes that AI must serve humanity beyond mere automation of routine tasks. An avid reader of books, Patrick lives in the San Francisco Bay Area with his wife and two children.

ACKNOWLEDGMENTS

This book is the result of many projects and experiences working with a variety of companies both as a product and services provider and an internal manager of AI teams. Many people have helped me over the years to learn and understand what is going on and what is needed to achieve a result.

Mentorship is so important in working life and I am grateful for the mentors who shared their time and advice with me. Thank you to Jeffrey Pfeffer from Stanford, Christian Anschuetz from Gartner, Reto Francioni from Deutsche Boerse, and Heinrich Harrer.

My managers have always provided me with many learning opportunities that have made me better over the years. Thank you to Scott Koo, Sean Im, and Shaun Lee from Samsung, Hardik Parekh from Searce, and Jeff Simmons from Oxy.

My coworkers are the engine of all the work I have had the privilege to be a part of. There are too many to name you all. Thank you for walking with me on the journey for a little while and sharing your energy. In particular, thank you to Benjamin Eckenfels and Markus Ahorner from algorithmica; Sima Didari and Dan Waters from Samsung; Kayla Spiess, Paul Pallath, Krati Doshi, and Mahek Bhatia from Searce; and Sabina Azizli, Nikolaos Mitsakos, Omar Riera, and Dipti Sankpal from Oxy.

Community is the foundation for progress. Without the ability to exchange ideas and receive guidance from peers in an atmosphere of mutual support and respect, most progress would not happen. Together we are more. The communities created by Evanta, now Gartner C-level Communities, and AIM Research have been pivotal to the creation and realization of many ideas in these pages. Thank you very much to my fellows. In particular, thank you very much to all those who took the time to read this manuscript prior to publication and wrote a review for it in the first pages and on the cover of this book.

Thank you to my family, and most especially my wife, Fahmida, for all the advice over the years and the time it takes to do this work—not just writing these notes but the real work that gave rise to them.

Introduction

"There is no plan B. Plan B is to succeed at plan A."

ARNOLD SCHWARZENEGGER, *BE USEFUL*

This is a handbook for all those who wish to lead a team of specialists using artificial intelligence (AI) to create real and lasting business value in an enterprise. It is not a book on AI neither as a scientific branch of mathematics nor as a technology topic in software form. It is more a book on leadership in the concrete context of AI.

AI projects fail to deliver business value mainly for three reasons:

1 Users do not want to use the software application that brings the AI model to them for a variety of reasons.

2 Users want to use it but cannot efficiently or effectively do so because the interface, either to the user or to the business process, is clunky in some way.

3 The AI application solves a problem no one has or rather solves only part of the real problem or a different version of it.

In essence, this book is a manual on how to avoid these three leading causes of death of AI projects. While avoiding them will not guarantee success, it will improve your odds by a factor of ten in individual projects.

In addition to the view on individual projects, this book discusses the reality of an enterprise AI program and team having a portfolio of projects with overlapping technological requirements and conflicting priorities. As your overall enterprise goal is not necessarily success in every project but rather the overall success of the AI program consisting of many projects, success is now virtually assured.

The root principle is a relentless focus on business value. AI should not be pursued because it is flashy, modern, or the latest hype but used surgically where it makes sense as a tool to solve a real problem.

Value can be achieved only if users use the product you build. The product needs to be fully integrated into the wider process for this to work and you need to build trust among the user group. Trust on the one hand is user confidence in the output of the AI application and on the other hand it is trust that the goal of the AI program is not laying off the very users you are dealing with. For this, the AI team must develop strong change management, empathy, and training skills.

AI scientists want to improve their models and hit accuracy records. While that is good, done is better than perfect. The AI team must embody standards for when an outcome is good enough and then move on because the goal is not scientific excellence but business value.

The key insights of the book are these:

1 Start with a clear, transparent, and concrete vision of what the AI program will deliver for the whole company—an enterprise AI strategy.

2 Structure a team and an operating model that can deliver on that strategy.

3 Identify use cases by looking for challenges currently faced by the business that can be solved meaningfully by AI and assemble these into a prioritized portfolio.

4 Create a comprehensive infrastructure for the deployment of the required AI models for the project portfolio using as many synergies as possible and taking into account all dependencies such as data access and quality as well as cybersecurity. The goal is not to deploy one use case after another but to deploy the entire portfolio!

5 For each project, carefully analyze business needs, write them down, and get business stakeholders to formally sign off on them. Use agile project management to conduct the projects with frequent stakeholder involvement.

6 Focus relentlessly on user experience and change management as these two elements together are necessary and sufficient conditions for successful adoption and therefore value generation.

7 Consistently emphasize the goal of value generation by carefully measuring cost and benefit throughout the project and watching project risks such as data quality.

8 Institutionalize central AI governance processes to ensure ethical and responsible use of AI throughout with appropriate risk management and guardrails.

9 Centralize the management of AI vendors such as research institutions, product vendors, and services companies to limit duplication of efforts and ensure high-quality standards.

10 Finally, create an enterprise-wide AI community through events and education fostering mutual learning and innovation.

In Part 1 of the book, team structure is discussed. Chapter 1 presents the main choices for an operating model of an enterprise AI team and suggests a hub-and-spoke approach as optimal. Creating an AI strategy and a group charter is next, which are essential for defining what this new team is going to do and not do.

Chapter 2 addresses all the people in the enterprise involved with AI but not part of the central AI team. This is the community of practice. One of the main functions of the community is to relativize the hype on AI that people are constantly bombarded with, especially by startups. Educating the community is an important aspect for preventing the central function from being overwhelmed.

Chapter 3 helps in identifying and prioritizing all the different use cases in the enterprise. Practical methods for discovery and ranking are presented that will help you to generate a roadmap of projects.

Chapter 4 points out that these use cases share many technological aspects. It is therefore expedient for the enterprise not to work on each use case in isolation but rather to develop a modular platform strategy into which the use cases will fit. This can reduce the cost per use case significantly over a portfolio.

Chapter 5 raises some risk awareness for the likely causes of project delay or cost overruns. It also discusses the strong need to place all projects onto a portfolio roadmap looking at time and personnel resources to get them all done.

In Part 2 of the book, we embed the AI program into the value stream of the enterprise. These chapters deal with individual projects. Chapter 6 presents design thinking as the method of choice for clearing up what a project is meant to do in advance. Aligning all stakeholders on a common and clear vision of the final state is, in my opinion, the single most important aspect of AI program management.

Chapter 7 discusses project management using the agile approach. This approach is almost the same as with normal software development but somewhat different due to the unpredictable nature of AI model training. As every project has some scientific risks, we will want some projects to fail out during the process, making sure that our project portfolio has a healthy mix of sure bets, reasonable investments, and risky shots.

Chapter 8 argues that there are no unhappy users. Users who become unhappy with your application will stop being users. A strong emphasis on user experience is a necessary condition for software, whether AI or not, to be accepted and acceptance is necessary for value generation.

Chapter 9 discusses change management as an essential element to generate trust and adoption among the user base. Both Chapters 8 and 9 emphasize that AI projects are, from an effort point of view, largely software projects that must be managed accordingly. It is a great failure of AI programs to look at AI use cases as principally issues of modeling, data analysis, or science of sorts. Enterprise AI programs are in the business of shipping usable products to real users.

Chapter 10 goes into the business case of a project with its costs and benefits. This leads to an overall investment thesis into projects for consideration strategically in the enterprise.

In Part 3 of the book, the dependencies on other groups inside and outside the enterprise are discussed. In Chapter 11, the necessity of good quality data is emphasized alongside methods to restore high quality, to establish data products with data owners. On a data science level, the idea of a distribution of residuals is raised as the key method for assessing model accuracy beyond single numbers.

Chapter 12 discusses ethical and responsible AI. Due to a lack of international agreement, the need for the enterprise to define clearly what it means by these terms is raised and some thoughts are generated on how to produce an ethical codex. Some suggestions are made on how to operationalize responsibility in general.

Chapter 13 deals with AI governance and introduces a comprehensive governance model for the whole life cycle of AI.

Chapter 14 considers the relationship between the enterprise and various external entities such as universities, product vendors, and services vendors who can augment your team's tools and capabilities. Suggesting buying what you can and only building what you must, maintaining good relationships with a small number of selected companies is highly recommended.

Chapter 15 emphasizes that AI is never done, and everyone involved must keep up to date. Particularly in AI, this is challenging, and so some suggestions are made on how to train leaders and everyone else in your enterprise.

Every section in the book begins with a quote. Many of the sources of those quotes are books that I recommend for further reading, mainly on leadership, that are listed at the end of the book. These quotes themselves raise excellent ideas that merit some thought.

This book is the first version of my comprehensive suggestions on how to build an enterprise AI program. As I learn new things every day, I would be most grateful to you for your feedback on how you found these suggestions, which worked for you, and which did not and what you did instead. Thank you very much!

PART ONE

The Optimal Structure of the Team for Success

1

Setting Up the Right Operating Model

AI Strategy

"That is what a clear vision gives you: a way to decipher whether a decision is good or bad for you. ... The happiest and most successful people in the world do everything in their power to avoid bad decisions that confuse matters and drag them away from their goals. Instead, they focus on making choices that bring clarity to their vision and bring them closer to achieving it."

ARNOLD SCHWARZENEGGER, *BE USEFUL*

For the purposes of this book, when I talk about an **enterprise**, I think of a company that has multiple business units, is in multiple locations, and has a number of employees that is larger than the number of people you can realistically interact with in a one-to-one capacity. This enterprise is likely to have several shared central service departments like human resources, finance, accounting, legal, reporting and so on that serve all the business units. Each of these shared departments may have staff members embedded in the business units, or they may act as a separate central team.

This is the context I will be assuming for this book in which I will describe how to set up and lead enterprise AI programs. Implicit in this is also the assumption that the enterprise is not, in itself, an AI company but most likely a company whose main business has nothing directly to do with AI or software. I am explicitly addressing enterprises from all industries, including, but not limited to, energy, manufacturing, transportation, logistics, household and consumer goods, banking, and media.

So, your company has decided to pursue AI. Congratulations! Why is it pursuing AI? What does it want to achieve and when? How will it go about doing this work? Who will be doing it?

The AI strategy's purpose is to provide a high-level guide to these questions. The items then need to be refined into a plan that will ultimately lead to tactical actions and timelines. Let's look at each question in turn.

Why is it pursuing AI? Hopefully the answer is not because it is a hyped topic, or because some management consultant told you to. If the honest answer is something like this, it pays to stop and think carefully whether there might be a better reason. AI is a toolset that can accomplish certain ends and not others. Starting early to think about what your needs are and whether AI is the right toolset to meet them is a great practice.

A phrase commonly used in this context is that AI is expected to provide "actionable insights." That sounds good. In thinking about what this might mean in practice, however, we quickly ask ourselves if we are truly prepared to do whatever the AI is going to end up telling us to do in a daily life scenario. Actionable insights will, after all, boil down to the AI making decisions even if there is a human in the loop who approves those decisions. I know of many cases in which humans ignored the actionable insights, so no value was achieved. I also know of cases where virtually every actionable insight was approved and some of those led to damage. In a hectic day full of events, human oversight may not be as effective as we think when we come up with a nice strategy. So, when you come up with an answer to why you are doing AI, please keep in mind whether that is a realistic thing to want.

Instead, here are some reasons for why you might want to use AI:

- You want to plan your inventory purchases and make use of price evolution to buy cheap. So, you create price forecasting models and follow their predictions even though they are sometimes inaccurate.

- You want to lower the rate at which your factory produces scrap and so develop computer vision models to detect scrap, as well as causal models to track the cause for scrap production. You then either reengineer the factory or use real-time AI models to guide the factory to avoid those causes. The same approach works for a retail company that wants to increase the rate at which people coming to a store end up purchasing something.

Ultimately, your reason for wanting to do AI should be because you want to accomplish something tangible and you are prepared to act on whatever AI tells you, knowing that this action will cost effort, time, and money to execute.

Any good strategy has desired outcomes and a time frame. AI applications may, for example, provide a new business model, revenue stream, cost savings, efficiency gains, automation, novel insights, or even worker productivity improvements. You might be thinking about a few surgical point solutions to improve specific bottlenecks, or you might be thinking about an all-encompassing workflow.

A frequent expectation is that achievements can be had quickly, with many plans promising an outcome in weeks. Those claims are true within limits that are usually written in the fine print of the plan and so not at all obvious to most business stakeholders. Quick results can be achieved when two main conditions are true. One, the problem is similar to a problem that the people or tool concerned have previously solved. Two, the result is a validation that this problem in principle can be solved in that way. This is often called proof-of-concept. Achieving stable results under real-life circumstances in daily operations almost certainly requires significantly more work over a longer time frame of months to establish a reliable data flow and deal with corner cases, which are atypical but realistic datapoints that may need special treatment due to their idiosyncrasies.

The quickest way to failure in AI is to set unrealistically short time frames and then to move on to the next project each time before the team has had a chance to reach stability. This is known as **pilot purgatory** where an enterprise continuously does proofs-of-concept and achieves scientific success but never deploys anything in productive daily use and therefore reaps no value. It is more common than you would like to believe. In fact, this is the normal state for many enterprises.

How will it go about doing this work? The strategy now needs a mechanism for achieving it. Part of the mechanism will need to be a partner ecosystem of product and services vendors. First and foremost, among these are cloud service providers who offer not only hardware rentals on the internet but a whole ecosystem of native software helping you to make and deploy AI. Workflows with existing tools can quickly be assembled by service companies to allow you to focus on proprietary matters such as your data.

Another part of the mechanism is the basic choices about your attitude to the buy versus build complex. The extent to which you want to build your own models based on your own data will necessitate creating teams and skills you may not currently have. Buying ready-made models and tools from elsewhere may be imperfect for you, but such models and tools are immediately operational.

Change management is an essential part of the mechanism by which you will need to socialize the AI journey with your workforce, and perhaps even suppliers and customers, so that an application is used and adopted by them. Only an application in use can possibly provide value.

Who will be doing it? Someone needs to be responsible and accountable. Is this a centralized team or a collection of decentralized teams? Perhaps a central team will own the strategy and plan, but decentralized teams will own the execution, such as in a hub-and-spoke model. We will discuss these options in the next section on the operating model. Together, these elements create an AI strategy.

Creating a strategy like this for an enterprise and achieving widespread understanding and agreement on it is a major project in itself. Even after agreement at the C-level has been reached, it will need to be communicated to every level of the enterprise. Many questions will be asked either about the strategy itself or about a multitude of details that are not directly in the strategy but will need to be known when getting to a plan of execution. These may include who will do what, how much it will cost, what technologies and vendors will be used, which AI model types will be chosen, what accuracies are desired. Such communication and question answering will also need to be done continuously over time as people forget about it or advance in their understanding of AI.

AI is a cross-functional technology. As such, the AI team will encounter virtually every department in the enterprise. They will want to know what the enterprise is doing and how they fit into the grander whole of the puzzle. The AI strategy is the picture of the whole, just like the image on the cardboard box that contains many puzzle pieces that must be assembled into that image. The strategy needs to be clear enough that everyone can recognize the big picture with some clarity and recognize their place in the vision.

I cannot overemphasize the absolute necessity of having a **clear vision** of why, what, when, how, and with whom you want to achieve some outcome. How can you know that you are getting closer or that you have reached your goal if you do not know what that goal looks like? In the daily trenches of doing AI, you will be distracted by many people, requests, scientific and business obstacles so that it is easy to lose your way.

This is especially true if you are responsible for either creating or communicating that vision and strategy around the enterprise or to partners like suppliers or customers. The ideal clear vision is, in fact, a vision … if you can draw a diagram or a small number of diagrams to communicate the vision, that is most helpful. We will see a few such diagrams in the remainder of this book that could form the basis for your strategy, suitably adapted to your enterprise, of course.

The Organization

"If you want to accomplish important tasks under trying conditions, you need to work with people the way they are, not as you wish them to be. You do this by learning to manage structure, not behavior."

MARVIN WEISBORD AND SANDRA JANOFF,
DON'T JUST DO SOMETHING, STAND THERE!

So, how should AI be organized at your company? There are several major options available and they depend on what you expect AI to do for the company.

There could be **one central AI team** that oversees any and all AI matters, with everyone else at the company being a consumer of their products and services. The advantage is that this group has full control and can thus probably achieve a secure and cost-effective service at the price of flexibility. The risk is that the people at this center will be far from the business and may release services that are not as relevant as they might be. Depending on the funding model, there is also a risk that many needs of the business will not be met as this team is overwhelmed by requests.

There could be **many small, decentralized AI teams** in each of the business units and central functions focused on applying AI to their sector. This model has the advantage that these teams are very close to the business and will solve real problems. It is likely to come at the cost of duplication as two groups with a similar problem may purchase essentially the same product or service independently of each other, increasing the cost of acquisition and maintenance for the company. It often also comes at the cost of technical expertise as decentralized teams will be tempted to hire staff who are not deep AI experts but rather domain experts who have been upskilled in AI. This is the model preferred by consulting companies as this offers ripe pickings for selling services.

The **hub-and-spoke model** is a combination of these two extremes. A central team would be in charge of the AI topics that are relevant to the entire enterprise, such as prioritizing use cases, selecting vendors, staying up to date with products, providing deep technical expertise, and solving mathematical or technical problems. The central team would also identify common elements between use cases where the company would benefit from creating or purchasing a central toolbox or service for all.

In addition to the central team, the embedded teams in the business units and shared functions (satellite teams) would focus on defining the challenges they face that can and should be solved by AI and supplying crucial domain expertise to make sure that AI solves the actual business problem and offers real value to the company.

As with the data organization or IT, these models have received a lot of attention and experimentation. Some enterprises have switched between centralized and decentralized several times depending on current executive leadership opinions. It is currently the general opinion among data, IT, and AI professionals that the hub-and-spoke model is the most durable and likely to create value long-term.

It is important to recognize the valuable contributions of all concerned and not to generate a competitive culture in which there is a two-class society between the center and its satellites. This model works only if everyone understands that they are part of a larger whole and only by working together can challenges be solved. The optimal combination of domain expertise and reality checking from the satellites and technical expertise and vendor coordination from the center can accomplish a lot of value.

A variant on the hub-and-spoke model is the **service model**. The AI team is organized just like a service company would be, only it is an internal organization to the enterprise and so does not require contracting relationships. The AI service organization has people who manage the relationship to the other departments and groups at the enterprise akin to the business development organization at a service company. It would have project managers to see the individual projects from inception to conclusion. There are enterprises in which this central services department charges the other departments money for its services and there are enterprises where this central services department is free of charge and funded by central budgets. Regardless of the flow of money, such an organization works like a semi-independent company.

In choosing which model to go for, it is expedient to think about the processes by which projects are prioritized for the internal AI team, discarded altogether, or outsourced to some external service provider or product vendor. A coherent enterprise-wide strategy and criteria are helpful alongside a central committee that has authority to make these choices. That committee may be in the central AI team, or it may include other business stakeholders. Including other business stakeholders is the better option as this provides credibility to choices made because the AI team is, after all, at the service of the enterprise.

To provide an example, here are some of the high-level criteria that you may consider implementing:

1 Use cases that offer additional revenue are better than those that save costs.

2 If you can buy it, do so. Build only what you cannot buy.

3 If it is worth less than $10 million per year, it will not be done. Alternatively, a use case must provide at least ten times its development cost in yearly value.

4 Risks will be evaluated according to worst-case impact and likelihood. Use cases may be rejected if they are considered high risk.

Relationship to IT and Data Organizations

"Resistance is so strong because, across the business, people do not believe that data and models are there to help them. They believe that data and models are here to help them end their careers. … People view AI as a teammate, not a tool. A new teammate replaces someone."

VIN VASHISHTA, *FROM DATA TO PROFIT*

The AI group, however it is organized, has two special relationships to the organizations dealing with IT and data. If you compare the AI group to a car taking the company from one place to another, data is the fuel and IT is the engine.

The IT organization will oversee the computational resources of the company, whether on-premises or in the cloud, which are crucial both for making and using AI models and tools. Cybersecurity is essential in making sure that only the right people have access and that no malicious services can interfere. The software ecosystem that AI gets data from, uses for processing, and outputs its results to is all likely managed and procured by IT.

Data governance, quality, storage, and access are likely controlled by a data organization that may or may not be a part of IT. AI must have access to high-quality data both for creating new models and for using its model for company benefit. The data must both be available and of high quality, otherwise all AI efforts are doomed to failure.

In the most positive scenario, having bad data is like trying to put gasoline into an electric car. You cannot do it and so you cause little harm apart from wasted time. In a more common scenario, it is like putting diesel into a regular car. The car is totaled, and significant value is lost—both the effort made to create the system and the potential rewards of its future usage.

These two special relationships are important and so some effort must be made in defining and establishing them so that a common understanding of goals and measures is achieved. The mutual relationships between the IT, data, and AI leaders are critical in this.

The ways in which these relationships are organized often reflect the nature of the company's business as well as its view on what value data and AI represent. In most companies, IT is seen as a cost center dominated by its duty to "keep the lights on," keep everything cyber-secure, and to do this with a minimal budget.

The view of data is shifting from being a technical issue of keeping data secure, properly stored, and accessible with low latency, to a function that adds value by connecting data silos with master data management, data catalogs, documentation, and data governance. The eventual vision of the data function is to create and own data products that live on the corporate balance sheet as bookable assets.

In contrast, the AI function should be creating value, preferably on the revenue side of the house. In creating value, AI products need to be integrated into established processes around the company, which are partly human, partly software, and partly hardware.

In most companies, the Chief Data Officer (CDO) and Chief AI Officer (CAIO) both report to the Chief Information Officer (CIO) as both data and AI are regarded as being aspects of IT. This view is true insofar as both ultimately exist in the form of software and are run on IT infrastructure. In terms of the company's business, however, this is not true. IT is a cost-saving function, data is an asset generator, and AI is a revenue generator. However, all three are cross-functional in that they do not belong to any business unit or region but need to work with them all and standardize their efforts across the company.

Another common model is to see AI as a technology and so have the CAIO report to the Chief Technology Officer (CTO) while data is a form of information so that the CDO reports to the CIO.

AI doubtlessly takes the form of software. The question is whether it *is* software or not. In its core, AI is not software but is the distillation of data into patterns that allow for interpolation and extrapolation beyond the data used to create it. As the data is generally about people, events, machines, and other elements of the world, AI is a representation of the dynamic behaviors in the world. You create it using scientific methods and use it for a practical purpose—to affect the world in some profitable manner. It is a technology. Therefore, I recommend having the CAIO report to the CTO. Alternatively, as AI is a strategic cross-functional division of the company comparable to finance, the CAIO could report directly to the CEO of your enterprise.

Literacy and Governance

"Models built in labs work only in the lab environment. The real world has completely different conditions, and customers have different expectations."

VIN VASHISHTA, FROM DATA TO PROFIT

An enterprise AI program will need to upskill and train many people around the enterprise. At a minimum, they need to be told what the strategy is and what is being done. It is better to educate the workforce on the basic principles of AI and how they are relevant to the enterprise and its various departments. You may choose to have four tiers of internal training.

First, the leaders of the enterprise want to know about the AI strategy and about the organizational and economic aspects of AI. They want to be able to recognize and differentiate AI from non-AI and want to be able to spot reasonable products and vendors. Aware that there is a lot of hype around AI, leaders are keen to understand what is real, what is a stretch, and what is hypothetical. The AI team, meanwhile, wants leaders to have realistic expectations both on the quality of AI products and on timelines and budgets needed to make them.

Second, many individuals will be interested in AI at a conceptual level. They will want to understand what AI is about, such as how to train a model and select a model type. Concepts like hyper-parameters, algorithms, and multiple model types are taught to this audience. Assessing whether a model is good or a dataset is representative are important basic concepts. Low-code no-code tools can be introduced at this level to help people get some initial contact with these methods and try things out.

Third, a smaller group wants to learn how to do AI themselves. These individuals will want to interact more seriously with the low-code no-code tool, and they want to learn the basics of programming languages to be able to create more intricate models of their own. The basics of SQL and Python are often taught here alongside some popular frameworks like PyTorch that enable the training of sophisticated models in a few lines of code. People in this group are often called citizen data scientists.

Fourth, an advanced group will now dive in deeply and learn how to conduct an AI project end to end. This group aims to do AI as part of their jobs and will join one of the decentralized satellite groups or the central AI team.

Improving the AI literacy of the workforce in general on these four levels helps the execution of a comprehensive AI strategy significantly as they will understand what you are doing and the context of why you are doing it.

While training courses like these are available online through training portals and external consultancies can be hired to perform training for your enterprise, I recommend that you do such training in house, in person, with staff from the central AI team. While the scientific AI content is, of course, general, a lot of the other content is specific to your enterprise. The use case examples, the location and nature of your data, the AI strategy, the software landscape, and the business ramifications are all custom to your organization. Additionally, courses like this develop a relationship between the AI team and the rest of the organization that will serve you well during execution.

Developing AI literacy is an investment in AI project success. It also acts as the foundation for AI governance. While governance itself consists mainly of various rules and processes of what to do and not to do, and how to do them or not to do them, it relies on people knowing these processes and following them. A widespread AI training course is the ideal medium for informing people about what they can and should do, what they had better avoid, and who to call in cases of uncertainty.

AI governance at a high level is the process by which we decide to buy a product, hire a service vendor, build a model or tool in house, pursue a use case, approve a model or tool for productive use, and retract a model or tool from productive use. We will discuss more specialized topics like responsible AI and ethical AI in Chapter 12 and model governance in Chapter 13.

This process needs to be designed at the enterprise level because many of these decisions will be relevant to the entire enterprise. It also needs to be realistic and lean because these decisions, in some form, will occur quite often. In larger enterprises, the committee conducting these processes may be asked for a decision more than once per day on average as it has been with the companies I have worked for. That practical fact means that the committee cannot be large, and it cannot include highly placed people, for example C-level leaders. Authority and responsibility must be delegated to this committee.

The frequency of decision-making also means that this committee cannot conduct an in-depth review of most cases and so needs to adopt some basic rules of engagement whereby easy cases are quickly approved or disapproved.

Based on this, it is clear that the governance process benefits from the larger population in the enterprise understanding the process and its necessity. This is best done in a training course.

Technology and Engineering

"In science if you know what you are doing you should not be doing it. In engineering if you do not know what you are doing you should not be doing it."

RICHARD HAMMING, *THE ART OF DOING SCIENCE AND ENGINEERING*

There are a great many software libraries and applications on the market to help with the preprocessing of data, the creating and using of AI models, and the postprocessing of the model result, ultimately ending in a graphical user interface for the human user or a data interface for a machine user. It is highly unlikely that you will be building any AI applications from scratch, but rather likely that you will be using a multitude of such components from a variety of vendors.

In fact, the availability of such components is so great that you cannot attain a full overview of them at any one time. Without great in-depth study, many tools remain virtually indistinguishable, and you will not have the time or resources to engage in in-depth studies most of the time. We will talk about the relationships with vendors in Chapter 14.

At the basic level of organization, you should make some major choices.

Hosting AI Technology

The most impactful choice is the host for your entire software landscape for AI. At the highest level, the four choices are to host this in one of the clouds, in an on-premises data center, in edge devices, or a combination thereof.

The modern default choice is, of course, a cloud. It makes little sense to compare the specific clouds here as they evolve quickly and have so many features that a comparison is a significant study in its own right. Every cloud comes with a set of software systems that are well integrated with each other, so joining a cloud means joining the ecosystem of tools available in that cloud. Signing a cloud contract is therefore not buying one thing; it is buying a large package of goods. Many of the goods in that package are positive for you and some may not be, as with every package. Of course, you can buy third-party software in any cloud to overcome any of these deficiencies.

For the same reason, once you have built out your AI landscape on a cloud, switching to another cloud can be quite difficult. The reality of vendor lock-in is very real, especially in the case of a cloud provider.

An on-premises data center is no longer a popular choice because it represents a significant capital investment at the start and because the maintenance of the hardware and software systems falls on your own staff. It is possible to rent a data center from another provider and so attain a half-way point between your own data center and a cloud. The difference between data center and cloud comes in three major categories. First, a data center is primarily constituted of hardware. You must supply the software yourself. A cloud comes with a host of software available natively. Second, a cloud allows you to dynamically scale up and down the amount of resources you use, while in a data center, you book a certain amount of resources for the long term. Third, a cloud includes comprehensive human and software services and availability of all hardware and software resources while a data center does not.

What all this amounts to is that a cloud is offering a comprehensive package with a 'no worries' guarantee, at a premium price. A data center offers a bargain price—after a heavy investment—at the cost of having to manage it yourself and the risk of making a mistake in estimating your needs.

An edge device is a different animal. These are hardware devices sitting close to the final application of the AI model. For example, AI models for driving cars should run inside the car itself and not rely on an internet connection. Industrial machinery like pumps and compressors need time-sensitive models and so they should run on a device right next to the machine. When the time from acquiring the input data for the AI model to the provision of the AI model output must be very short—such as tens of milliseconds—it called a **low latency** application. These applications, like voice translation into foreign languages, should run on your mobile phone and not be looped through the cloud to prevent an annoying or prohibitive time delay.

If you need edge devices, you have many hardware options that change frequently. The strategic point however is that you will quickly end up with a fleet of these devices that is likely to be deployed over a wide geographical area. Maintaining the fleet is the challenge.

The best way to handle this is for the edge devices to periodically check, via the internet, with some central service whether there is an available update to the model and to automatically obtain and download it. That central service can run in a data center or a cloud, of course. That is not enough, however. The edge devices must also upload a selection of their data periodically. This is to check whether the model is still good enough and to provide fresh training data to the central service for periodic model retraining when the model is no longer good enough. We will discuss the complexity of AI observability in Chapter 13.

In reality, for most enterprises, it will be a combination of these options. You are most likely to do the bulk of the work in the cloud, with edge devices involved for real-time work or for places that have limited internet connectivity. While the on-premises data center is increasingly less common, it does make sense for the workloads that are niche and constant over longer periods of time as the cost of the cloud is higher and its benefits derive from flexibility and interoperability.

Data Input Pipelines

The next strategic decision is how to get the data from its source to the AI models. The source may be a physical sensor device measuring some quantity like a temperature, it may be a camera streaming a video feed, it may be a microphone streaming an audio signal, or it could be a human typing text or creating some file. This data must be collected from the source, converted into a uniform format, and stored somewhere for processing by AI.

In the case of a cloud or data center, this data storage location will be a central location. It is most likely a data warehouse for structured data or a data lake for unstructured data. Making sure that the source data arrives at this central location in a timely manner without getting dropped is a complex endeavor. Some forms of data such as images or files will need meta-data to be stored in a database so that they can be administered. Retention policies will need to be created to make sure that you keep enough data for future model training and refinement but not too much to exceed your budget for data storage—this is especially relevant for image or video data that can quickly consume large storage volumes.

Incoming data may need to be reduced at the earliest opportunity to keep the size of the data traffic manageable across your computer network. Audio and visual signals especially generate data that could clog up the network. Those signals would need to be reduced on the device that acquires them, i.e., the microphone or camera itself.

Being able to access the data with relatively low latency whenever needed will require storage policies in several tiers to be implemented. We may have hot storage that is kept in memory for immediate retrieval, warm storage that is kept in solid-state drives for quick retrieval, cold storage kept on rotating drives, and glacial storage kept on tape drives just to provide a basic overview. Most clouds have a much more refined offering of storage tiers available.

Data quality is of primary significance to AI. If the data is of poor quality, AI will fail. Beyond assuring that the data gets captured and stored properly, rules and systems are needed to monitor data quality and, possibly, improve or restore it. While this can get technical and intricate quickly, conceptual understanding of this is important and should be discussed. A few popular examples are treated here to provide a starting point.

In measuring a time-series such as a temperature that is recorded every five minutes, we may have a temporary sensor failure or connectivity problem leading to several hours of missing data. AI does not deal well with missing data so the AI model will either not work for this time or you will need to fill in the data in some way. Most of the industrial systems have decided that the last measured data point will remain valid until a new data point is measured. That is a decision to improve the data quality, but you could choose differently.

Sensors drift over time as they get old, damaged, dirty, or somehow compromised. A value may therefore slowly increase or decrease over months. Gradually, therefore, the values are wrong. When replaced with a new sensor, the values appear to jump. To the AI that signals a qualitative change. Nothing has changed in the real situation—only the sensor has changed. This is a data quality issue. You can detect such drifting with AI, but this represents an extra effort in creating and deploying the AI models that assure good data quality for the actual AI model you are interested in.

Your dataset may have duplicates in it. This wastes space and computational resources. It also overemphasizes that case. Depending on how many times a data point is duplicated, this may not be a big concern. For retail use cases, for example, having a single real human being represented as many customers in a database could potentially be quite problematic and costly.

These duplicates must be found and removed. However, they may be hard to spot because they are likely not identical in the database and may differ by some trivial amount, such as a spelling error or an old address.

Another interesting problem of quality is units. There is often some confusion over whether a length is measured in feet, yards, or meters. A temperature could be in degrees Celsius or Fahrenheit. Even the order of magnitude may be under dispute, whether it is a kilo-, milli-, or micro-unit for example.

Apart from the actual unit, it often happens that an important quantity is in the wrong units entirely. This happens particularly in formulating questions or challenges to be solved. For instance, it makes sense to talk about your monthly salary in currency terms, in dollars or euros. But to express your wealth, it makes more sense to measure that in time units rather than currency units. A great measure of your wealth is the amount of time you can live entirely on your wealth alone, without an income, and maintaining your current lifestyle.

AI Training Frameworks

Once the infrastructure and data are sorted out, the data science can begin. This usually consists of several steps, each of which can consist of significant work and sub-steps. In order, they are data curation and cleansing, data labeling, feature engineering, and model training. Each of these steps needs tools of its own.

Data curation and cleansing takes the data from the data warehouse and selects the section of data that will be used for this exercise. That data is prepared for model training by cleaning it. This consists of removing bad data points and perhaps adjusting some existing data points. The most common action is removing duplicates and outliers. Depending on the data's format and nature, this process can be easy or cumbersome. Sometimes the data warehouse offers tools to do this and sometimes an investment in data-wrangling tools is expedient.

Data labeling is an activity that adds human domain knowledge to the dataset, usually via direct manual input. This is most typical in computer vision use cases where human experts will use the computer mouse to draw rectangles or custom shapes over an image to indicate that this area of the image is the pattern that the AI should learn to detect, for example the cancer on a medical image. Due to human error, it is considered standard practice for at least three independent human labelers to label each data point and then to take a democratic approach to the decision on whose opinion to take. This is to say that there is more to labeling than merely recording inputs. This process alone can be responsible for much of the financial investment in an AI use case. There is a cottage industry of labeling service companies that will do this on your behalf. Such companies have the necessary tooling to do the job and can efficiently label a large number of data points.

Feature engineering is probably the most consequential step in terms of AI model accuracy at the end of the road. It consists of two sub-steps of feature generation and feature selection.

Feature generation is where human domain experts advise the data scientists as to what to look for and how to look for it. For example, a dataset might have columns A and B in it. The domain expert will point out that it is actually the ratio A/B that is important, or, more complex, that the moving average of A over a time-window of 17 minutes is especially informative about this special situation. In general, the aim at this stage is to introduce not a few but rather a great many new features into the dataset. It is not uncommon at this point to increase the number of columns in the dataset by a factor of up to 10, going from 10 raw features perhaps to 100 total features. This is the data scientific way of admitting that there is something that it is not known and that an open mind is being kept.

Feature selection is where the number of features is radically reduced by selecting those features that offer the most information about the problem at hand. This is done objectively based on various strategies. Simply, we can use correlation analysis to drop those that don't correlate with the desired answer or those that correlate with each other so highly as to effectively be the same feature. More intricately, we can perform analyses such as principal component analysis whereby the data points are projected into a smaller dimensional space while preserving maximal variance—this method will not be described here; suffice it to say that it makes sense to apply something like this nearly always. It must be said that at this time there is no known method, apart from trial and error, that can derive the smallest number of features from your dataset that offers the highest amount of information. This selection task remains more of an art than a science.

Model training follows in which the type of model must be selected by a human AI expert, the hyper-parameters of the model training algorithm are tuned, and the training actually executed. There is also no known method currently to derive the best possible hyper-parameters without doing a trial-and-error analysis, so this step is potentially time-consuming. This step concludes with the assessment of the model using test data at this time—data that the model has not seen up to this point. If the model is good enough upon critical examination, it can be deployed, a step we will take in the next sections.

All these steps need tooling to be done, as you might imagine. Of course, all of this can be done by writing source code, for example in Python, in a programming environment, for example a Jupyter Notebook. However, this approach is not robust, nor scalable. It is not robust because it is hard for another person to pick up where someone else left off, perhaps months ago. It is not scalable as this leads to a labyrinth of code files for a larger distributed team, particularly if you have a decentralized organization. After some amount of time and some number of developers, such code may effectively not exist for all the good it provides to the enterprise.

After the models and their software packaging with data input and output and user interfaces are all written and tested starts the process of maintenance. While not as glorious as the creation of the application, its maintenance is a significant process that could, over the years, easily consume many times the resources that it took to make the application. The model must be observed to detect when the accuracy decreases, for which an **observability platform** is needed. The software packaging must be maintained also, for which you need dedicated staff and regular communication with users.

A careful design of a small number of tools in which workflows can be created that rely on a minimum of custom code is the way to go. Many choices for so-called **orchestration platforms** exist on the market and make an excellent investment for an enterprise with a significant number of both (citizen) data scientists and use cases.

Model Approval Process

Once made, models must go through an approvals process to reach deployment. It is a best practice to do this via a specific central group so that all models undergo a rigorous and objective testing methodology. In no case should the model developers themselves be allowed to decide to deploy the model as they are too close to the project and have a vested interest in releasing it.

As with all good processes, this approvals process should be documented, models versioned, and all of this stored so that decisions can be traced. Should anything go wrong, this process then allows the deployment framework to quickly and efficiently default back to the most recent earlier version of the model. Unsurprisingly, there are platforms that manage this process well.

Deployment Methodologies

Having decided to deploy a model, it is time to actually deploy it. This is the big moment at which the model meets live data and is meant to be used by real users in a normal everyday environment. At this point, the model ceases to be a mere model and starts to be an application, together with the data input, preprocessing, postprocessing, output, and user interface layers. All of this must be protected by data encryption, in transit and at rest, user authentication, and cybersecurity. The computational infrastructure must scale up and down as necessary. The data must be streamed through the application at the right cadence. And so on. Deployment is complex!

In fact, the amount of effort to build a good deployment structure for an AI model can take several times the effort needed to make the model itself. Different skills are needed to do this, and this process requires a plan and budget to execute successfully. The relationship with IT comes into its own at this point as well because the AI team will need all the help it can get to productionize a model.

If your enterprise has a handful of models, you may choose to deploy each on its own merits. If you have industry-standard use cases for which you can effectively buy the deployment structure from a vendor, you should do so. However, if you are a large enterprise with more than a handful of use cases, some of which are niche to your particular circumstances, you will benefit greatly from a custom-designed deployment platform. Chapter 4 will go into this in more detail.

Creating a Charter for the AI Team

"So much of what data scientists do fails because you refuse to meet the business where it is and let it dictate the pace of transformation."
"Transformation is hard because you have users who have been lied to. You may not have lied to them, but they don't care."

VIN VASHISHTA, *FROM DATA TO PROFIT*

Now that you have a strategy, an operating model, a good relationship to IT, a way to generate AI literacy and governance, and some basic technologies decided on, take a deep breath. You have settled the why, what, and how. It is time to decide on the who. Then you have a charter for the AI team. The who part covers four main questions of authority.

1 Funding: Who pays for what parts of which projects?
2 Prioritization: Who decides what projects are approved or prioritized?
3 Management: Who manages projects and who maintains artefacts?
4 Responsibility: Who oversees ethical and responsible AI and regulatory compliance?

Funding the AI Team

The work of the AI team is in part organizational, infrastructural, and project-specific. These three areas are highly dependent and overlap. Much of the organizational and infrastructural work supports all use cases but cannot be tied to any one of them directly. The source of funding is often a difficult topic in an enterprise.

There are enterprises where the AI team charges real money to the business units for any work and products it provides and the AI team thus gets its own profit-and-loss statement, making it into a business unit in its own right. There are enterprises where the AI team is entirely funded from central budgets and is therefore completely free for the business units to use, with the possible exception of extra purchases that must be made in service of a specific use case, which can be charged to the customer business unit. There are also enterprises that have a hybrid funding model in which the business units pay for parts of the products and services provided by the AI team. This gets particularly complex in a decentralized operating model in which the satellite team is already a part of the business unit's budget.

While the particular funding model is up to your enterprise to decide, I strongly recommend that this be decided as early as possible and made extremely explicit to everyone concerned so that there is no confusion about the terms of engaging the team to prevent misunderstandings and wrangling after the fact.

Approving and Prioritizing Use Cases

Of the multitude of use cases that everyone around the enterprise will want to pursue, someone must decide which ones to engage in and which ones to drop. Projects that will be engaged in enter the roadmap and may have to wait for resources to free up—not every project can be worked on right now. The same decisions need to be made for all the AI products that people will want to buy. In a larger enterprise, these two decision processes may realistically occur once or twice every single business day, on average.

It is therefore important to have a transparent, objective, robust, and efficient process for making these decisions. For that, it must be clear who is responsible for this and for all of that to be communicated to everyone concerned, most especially to the staff of the procurement organization and the legal department.

Managing Projects

Managing AI projects is not easy. Chapter 7 will go into this in detail. Organizationally, project management requires dedicated project managers who are educated in the craft of making software products. Creating software is different from creating physical objects and so has its own approach, usually called agile. It is not ideal to have projects managed directly by the developers working on the projects, managed by the customer from the business unit, or having no formal project management at all and default to hope that the developers will eventually get there. These are all good ways to ensure failure.

The point of this is to say that since you need dedicated staff for agile project management, you will need a budget for it.

Responsibility for AI

The word 'responsibility' and AI are closely linked. The field of Responsible AI that grew out of Ethical AI or AI Ethics is a vibrant and significant area of academic research and industrial practice. As governments release regulations, this topic also starts to get into contact with compliance methods that seek to document adherence to these regulations, such as the AI Act in the European Union (EU). The global regulations for AI are currently in constant flux and largely nonexistent except for a few countries and so this space will see a lot of turbulence and growth over the next decades. We will cover responsible AI in Chapter 12.

From an organizational point of view, you must ask three root questions: What sort of use cases are okay or not okay to be pursued? Who is going to decide that? What checks and balances must we put into the process and the application?

In answering what is good or not, we need a clear and general definition of "good." The current best practice in Responsible AI is to define this based on a set of values or principles that define what the enterprise considers to be ethical, responsible, and good in its context. That is a journey of self-discovery that is not quick or easy and will need discussions at the highest level of the enterprise to reach a baseline that will stick.

The committee that decides on approving use cases and tools can also be in charge of conducting an ethical, responsible compliance review, or you may consider having a separate committee in charge of this. For most enterprises, the same committee will probably be good enough. In particularly sensitive industries where many use cases have direct impact on humans, a separate AI responsibility committee may be prudent.

KEY TAKEAWAYS

1 Create and communicate an enterprise-wide AI strategy.

2 Agree on an operating model and how the organization will upskill the enterprise, govern AI, and create infrastructure for AI.

3 Establish accountability for funding, approval, prioritizing, and being responsible for AI.

2

Building a Community
of Citizen Data Scientists

Data Science Is a Team Sport

*"Including all the relevant people in each meeting produced faster action
on problems, decisions, policies, and plans than any other strategy."*

MARVIN WEISBORD AND SANDRA JANOFF,
DON'T JUST DO SOMETHING, STAND THERE!

The AI team, whether centralized or not, cannot do its work alone. AI and data
science are cross-functional fields. Support will be needed at all levels of the
organization with varying degrees of intensity. Some members of the organization will be involved very closely in AI projects and may, for a while, spend most
of their time working directly with the AI team. For all those people, it makes
sense to create what I call a **community of practice** for AI. This chapter is about
what that means, why it makes sense, and some ideas on how to set it up. The
community of all those in the enterprise connected to AI projects include
several general groups.

First, it includes all those who directly write data scientific and AI code in
programming languages or create models in AI frameworks. These people will
be referred to as **data scientists** whether they work in a formal AI team or not.
Some of them may work in IT while others may work for the business units.
Their distinguishing feature is AI scientific and programming competence.

Second, it includes a variety of people who are experts in the fields that
the AI models are about and the AI applications will support. These are the
domain experts. They may provide specifications, advice, expertise, and will
collaborate with the data scientists principally by labeling data, designing
features, and ensuring model suitability for the practical purpose.

Third, it includes **management** at various levels in the hierarchy of the enterprise. Management provides the permission to use the data, the funds to pay for what is required, the prioritization of the applications that help the business achieve value by using the application.

Fourth, it includes various stakeholders who neither develop the application, nor weigh in on its details, nor provide management support. These stakeholders are most often the users of the application. Depending on the application, this may not be a homogenous group of people. They primarily interact with the other groups by specifying how the user interface and user experience must look to them. Ultimately this is the most crucial group of all because if the users do not use the application—for whatever reason—no value is generated for the enterprise.

A summary of these four groups with their interactions and roles can be found in Figure 2.1.

FIGURE 2.1 The interaction between the four groups in the community of practice

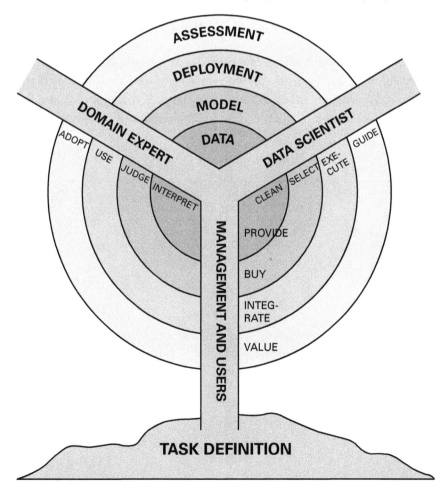

It will be clear from this description that the community of practice can potentially include a considerable portion of the enterprise's staff. The different groups will have different needs and so the community is not a singular entity but needs to be split into sub-groups as described.

The principal need for the community is training. Beyond that, the community benefits significantly from identifying itself as a community, learning from each other, and sharing information with each other. This information sharing is particularly helpful in this special case—the vast and quickly evolving landscape of startups and the consistent hype in AI. The next sections discuss these topics.

Training the Community

"Low trust causes friction, whether it is caused by unethical behavior or by ethical but incompetent behavior."
"When trust is low, speed goes down and cost goes up."

STEPHEN M.R. COVEY, *THE SPEED OF TRUST*

As this community is made up of a large number of people spread out over many different departments, the current AI knowledge will be similarly diverse. Not everyone needs to know everything. At a high level, Table 2.1 provides a suggestion of what the four main groups need to know.

TABLE 2.1 Training needs for different groups in the community

	Data Scientists	Domain Experts	Management	Users
Charter and Processes (Chapter 1)	X	X	X	X
AI Concepts	X	X	X	X
Responsible AI	X	X	X	X
Data Cleaning/Labeling	X	X	X	
Agile Project Management	X	X	X	
Effort/Cost/Value	X	X	X	
Feature Engineering	X	X		
Orchestration Tools	X	X		

(continued)

TABLE 2.1 (Continued)

	Data Scientists	Domain Experts	Management	Users
Deployment Frameworks	X			
AI Modeling and Science	X			
Programming Languages	X			

Everyone must be clearly informed about all the high-level decisions taken to form the charter of the AI program in the enterprise—all the topics that were discussed in Chapter 1. This is crucial because questions on those points will occur often, and doubt will create many problems.

Basic AI concepts should also be understood by everyone. Such concepts include knowing that AI models take many different forms but are all trained on data. No model is perfect and so a model providing a wrong answer is not a bug but an expected behavior that occurs—hopefully rarely—but will occur. Analyzing the practical ramifications of false positives and false negatives is important for all concerned. Everyone should understand how training and testing work and how to judge the goodness of a model—for instance by looking at the distribution plot of residuals, the differences between the model and the true answer.

Concepts of responsible AI and the enterprise's attitude to it—the values and principles—should be known by everyone so that they can assess their use cases and approaches in light of this framework.

How to conduct data cleansing and labeling or how to manage projects in the agile fashion as well as the economic aspects of estimating effort, timeline, cost, business value of AI need to be known only by the data scientists, domain experts, and management. Users do not need to go to this level of project detail.

Feature engineering and the usage of orchestration tools is now a technical matter that concerns data scientists and domain experts but no longer management.

Finally, deployment, AI modeling, and programming concern data scientists only.

Resources for this training should be created and courses made available widely in the enterprise to disseminate this information. It will then need to be reinforced in the projects as these groups start to work on real cases. It is now, in the work on real models, that all these theoretical concepts become practical and gain a life of their own. It is essential that the people managing the projects and overseeing the AI program live these concepts, practice them with everyone else, and remind people of them so that over time they become institutionalized. It would be a mistake, as usual, to think that merely informing people of these practices will lead to them being implemented. Training is therefore not an event but a continuous activity.

Community Building and Events

"Collective intelligence depends less on people's cognitive skills than their prosocial skills. The best teams have the most team players—people who excel at collaborating with others. … It's about figuring out what the group needs and enlisting everyone's contribution."

ADAM GRANT, *HIDDEN POTENTIAL*

All the people previously discussed recognize that there are many others at the same enterprise and at other enterprises who are trying to do similar things. Many of them will hear a lot about AI from coworkers, friends at other companies, family members, and, most importantly, vendors and the media.

Opinions will abound and will likely cover the full spectrum, from extreme optimism to extreme pessimism, about what AI is capable of, the changes it will bring, the timescales of those changes, and the threats it poses to the job market and to society at large.

The AI team will do well to be aware that this is going on and will benefit significantly from streamlining the conversation in curated events that generate a conversation about these issues. It is probably positive if the AI team does not dominate but merely organizes a forum of discussion, but it will do well to provide a voice of reason and scientifically backed point of view on some of the more extreme opinions that could be divisive.

You might consider holding regular presentations on AI projects. Given by a business stakeholder such as a domain expert, end user, or member of management, such presentations can provide an insightful look into a project for other non-experts in AI. They can talk about the context of the challenge, the approaches taken and discarded, the approach that was adopted, the practical changes that were adopted to make the AI solution work in real life, and ultimately the value achieved. Many presentations may cover only part of this if the projects are not finished yet.

Believe it or not, some of the most successful and insightful presentations are about projects or scientific attempts that failed. It would be a mistake to invite only successful projects to present. One can learn more from a failure than from several successes.

Several formats for such presentations could be envisioned. Longer presentations of an hour could go into considerable detail and offer scientific insights into AI technology and engineering. Shorter presentations of 20 minutes could focus mainly on the business challenge. For more detailed presentations, it is likely that visuals will play a role and so a webinar format, or even an in-person lecture, is the right setup. For a shorter presentation, visuals are less important and a podcast format might work best. As the enterprise generally has multiple locations, it is important to remember that this community is geographically spread out and not to divide the audience merely by location.

A format I have found to work well is one longer presentation every two weeks in a lecture hall environment at the head office that is streamed as a webinar as well. Such presentations alternate between a project presentation and an educational topic about AI as such. Those educational topics could cover new methods, models, tools, general insights, current affairs, regulations and more.

Internal Conferences and Workshops

"Faster and shorter often means more frequent, less conclusive meetings with costly unintended long-term consequences."

MARVIN WEISBORD AND SANDRA JANOFF,
DON'T JUST DO SOMETHING, STAND THERE!

Once your internal community reaches several hundred people, it may make sense to organize an internal gathering once or multiple times per year. One larger event per year could function in the same way that any other conference does. The big difference is that it is for internal participants only and so the presentations and conversations can include proprietary details and people can be more direct and honest about what worked and what did not.

One could have panels with senior managers, presentations by business stakeholders, technical talks by data scientists, and experience reports by end users.

Often the different business units of a company do not converse regularly and it takes an event by a cross-functional unit to generate a debate. AI is a function that cuts right across all divisions and so uniquely offers this possibility.

Events like this cost money and take effort to organize. Realizing this, you may be tempted to find sponsors. In my view, that would be a mistake because now external people would be invited, a sales pitch angle, however small, would be introduced, and participants would feel less free to debate.

Workshops are another great opportunity for community building. The words presentation and workshop are often used nearly interchangeably. What I mean by workshop in this context is an event where the participants do more than listen to a speaker but genuinely participate and co-create something as a group.

Creating elements of the AI strategy and the other major components of the AI charter discussed in Chapter 1 are great outcomes of workshops with the right senior participants. Agreeing together on a common problem statement, a common understanding of available data, and the design of informative features for a challenge are great examples of more scientific workshops. These can be conducted either to actually achieve those outcomes in a real scenario or to train participants in those concepts using a contrived simplified scenario.

Such events build community, enable collaboration, and encourage people to learn from each other. This will also lead to more convergent opinions on AI across the enterprise and to better adherence to policies and procedures as well as to better utilization of the available tools.

Keep the Startups and Hype at Bay

"A startup is a human institution designed to create a new product or service under conditions of extreme uncertainty."

ERIC RIES, *THE LEAN STARTUP*

Most companies active in the AI field are startups and much of what is said in public about AI is a form of hype. This section will provide some background information and strategies for dealing with startups and hype to help steer a good course through the noise and complexity.

How Do Startups Work?

There are many opinions on what a startup is or whether a specific company is a startup. It is not a well-defined term, nor is there good agreement. Philosophically, it can refer to a company that is—relatively—young, small, creating a new product or service, creating a new business model, creating a new market, or seeking rapid growth. More down to earth, it often refers to a company that receives funding from an investor because it still needs funding to survive. As such, this is a company that has not yet quite figured out how it will transform itself from a company into a business. Typically, it is also still iterating on its offering as it is figuring out how to become a business selling that offering.

As a principal feature of startups is funding, it is instructive to look at this process. Startups are usually founded by a team of three to five founders who have complementary skills in technology, business, finance, and marketing who go to venture capital companies in search of **seed funding** to get started. At this point, the team typically has little more than a great idea with a pitch deck. That idea is now built into a first offering ready to be presented to a select few test customers. This is the point where the company seeks what is called **Series A** funding to pay for the necessary incorporation of early feedback from the test customers. It is around the Series A time frame that the company, for the first time, makes any revenue from a customer. The startup graduates to **Series B** funding when a solid product-market-fit is achieved and the focus can shift onto acquiring lots of users or customers. This is known as scaling the company. When it is time to access new markets, create new products, acquire competitors, or prepare for a public offering, the company seeks **Series C** funding. Beyond this, there may be several more funding rounds to achieve ever greater stages of scaling. These rounds are meant to help the company achieve a specific set of targets acknowledging that the company could not achieve them on its own, or only slowly.

Many startups spend a long time in what is known as the **pre-revenue** phase. That is the phase in which the company has not received any customer payments. According to current philosophy that is heavily influenced by the "lean startup" idea from the book of the same title by Eric Ries, startups should focus on **validated learning** from early users to quickly get to a product that solves a real problem for real customers. In an attempt to do this, many startups forgo charging customers early on or at least charge them only nominal amounts.

From the point of view of startup founders, there are two markets. The market of customers paying fees for products or services is the secondary market. It is important for customers of startups to understand that the primary market is the collection of investors—mostly venture capital and private equity—who will want to buy the company from the current investor. There is a market for products and services and a market for entire companies. Founders and investors typically care about the market for companies. The point of sale of the company is known as the **exit**, and it is this that mostly motivates founders and investors.

Therefore, when startups talk about **growth**, they usually talk about the growth in the number of users (or a similar metric) and the speed of increase of such metrics. They are *not* usually talking about financial measures such as revenue or profit. Growing the company means growing the future expectations or valuation of the entire company to a future investor.

Once a company has turned into a business—meaning that it is turning a stable profit—that has lots of paying customers to keep happy and so is not primarily chasing fast growth in this sense, it has effectively stopped being a startup.

Pros and Cons of Working with Startups

With this context, an AI startup will seek to work with an enterprise for several possible reasons. First, it wants to have the enterprise as an early customer to provide validated learning to the startup or placate investors. Second, it wants to add either a good logo or many users to its user base in an attempt to generate growth. Third, it wants to use the enterprise's reputation and voice to break into the industry that this enterprise is in. Fourth, it wants to earn revenue from the enterprise.

Early-stage startups, before the Series B stage, will almost certainly want to generate validated learning. In fact, they may be sufficiently early that they will want to create their first real offering together with you. There is a popular concept of a **development partner**, which means that the startup will develop an almost custom product for you with the understanding that they can subsequently productize this and sell it to others. The startup will use your domain expertise and data to gain knowledge and insights into your industry.

The reason for pointing all this out is that a startup typically wants different things, or at least more things, than money from the enterprise. That fact should be understood when buying from or partnering with a startup.

If you are prepared to let them in and interact with your domain experts to learn the industry, they may reward you by building a custom AI toolset and models for you at a bargain price, sometimes at zero cost to you. The enterprise may even be able to get significant ownership stake in the startup if that is interesting to you.

The more mature a startup becomes, the less flexible it will be to customize its tools for you and the more it will seek a large customer base, and eventually will seek to maximize profitability. A startup is very malleable before Series B, will seek to expand customer numbers in Series B, and will want to become a stable player in your industry in Series C while it still cares about building reputation by being connected to an enterprise. Beyond that, it is likely to significantly seek revenue and profitability and start acting like a regular non-startup company.

The downside to all this flexibility is that startups change their minds quickly and greatly based on validated learning. This is known as **pivoting**. That is, in fact, the whole point of validated learning. If what they learn from you convinces the founder that this will eventually lead to a business with a high valuation, all is well. If not, the startup, or its investor, may choose to stop. If a startup fails to get the next funding round, it may no longer exist. Focusing on getting to a basic product as quickly as possible, the product is often raw and undocumented so that it may take a long time for the product to become stable and what one may call "enterprise grade." Much of the knowledge and ingenuity of a startup is contained in the brains of a handful of people. If they choose to stop doing this, the product effectively ceases to exist. These are risks and should be kept in mind.

All in all, cooperating with startups can be extremely rewarding for an enterprise by getting an independent, small, and highly motivated group of smart people to look into a problem on their behalf—often at a bargain price even.

There Are So Many Startups

The most frequent problem with startups is simply that there are so many of them. Every day, quite a number of them reach out and want attention. They all attempt to create fear, anxiety, greed, and most of all fear-of-missing-out (FOMO).

If a startup effectively says that they can do everything, delete the message. If they say they just want to learn what you care about or just have a discussion on AI, delete the message. If they are asking you to "jump on a quick call," delete the message. If they only want to show you some new technique that they have revolutionized, delete the message. If they present your own insights back to you and want to praise you to get a phone call, delete the message.

My advice is to only take them seriously if they can demonstrate some appreciation of the challenges you in your industry have and they can credibly claim to have made a dent into that challenge. A good marketing message ought to be able to do that in one or two sentences.

Even better, you might consider having a professional filter them for you by investing in a venture capital fund or accelerator program.

Regardless, you must come up with a robust and nearly impermeable filter for the marketing of the many startups. This is particularly relevant to AI since the number of companies is so large and it can be difficult to tell if they can do what they claim without putting them to the test, which you can only do for a very small number of them.

How to Spot Hype

The main consideration in that filter is to spot **hype**. To me, and in the context of AI, hype has two components. First, people exaggerate a capability by making its accuracy, relevance, effort, cost, timeline, or value seem better than it is. Second, capabilities are claimed that, at the current state-of-the-art, are not possible.

One rule of thumb is to critically investigate claims originating from people who personally stand to gain from the claim. As the conversation in AI is dominated by a small number of companies and prominent individuals at those companies, it is easy to see that this rule effectively eliminates a significant proportion of the conversation or at least puts it into doubt. Many claims you might hear may not come directly from those sources but they may still have originated there by word of mouth.

If independent experts in the field validate those claims, then they can be taken seriously. If they matter to your enterprise, you should test them for yourself. The main dimensions of hype mentioned above will now be discussed in turn.

ACCURACY

Claims that an AI model is more accurate than humans or has an accuracy close to 100 percent must be regarded with some suspicion. Other than an outright fabrication, it is likely to be true in certain circumstances and within certain well-defined boundaries. The boundaries may place restrictions on the values of input variables or have expectations on data quality. The marketing message may well have skipped them and they must be found. These boundaries may make the model irrelevant to your application or to the real-life circumstances of your enterprise.

RELEVANCE

An AI application that solves exactly this challenge may or may not be relevant to that other challenge, even if they are closely related. It depends primarily on two aspects. The AI model's relevance depends on whether the distribution of the model's training data matches the distribution of your data. The AI application's relevance depends on whether your processes—human and computerized—match with the expectations of the vendor. Both need effort to determine. Be critical in assessing the circumstances and context of the advertised application.

EFFORT

Claims that something can be done in a very short time are typically unrealistic. Startups especially often do not consider corporate processes or the time delays in communication. Short timelines often assume that the customer is a single individual who can do whatever is needed without having to ask for any permission. Short timelines in training or in achieving high accuracy often assume that nothing will go wrong or that the data is immediately available at a high volume and of pristine quality. Installing enterprise software in a day, training a model in a week, doing a proof-of-concept in a few days are simply not realistic outside of a contrived academic exercise.

COST

Vendors typically consider only the cost to them when they quote a price to you. Costs in third-party infrastructure, products, or services are usually excluded. Your internal costs in domain expertise, internal communication, IT, overhead, and change management are usually not even acknowledged. Beyond the mere project itself, software costs—and AI comes in the form of software—are principally in long-term maintenance and upkeep, which also usually go unmentioned. Startups especially may not even charge a realistic price for the project if they are trying to get into the market through you, further exaggerating the seemingly low price.

TIMELINE

Alongside the effort dimension, the project's timeline is often exaggerated and unrealistic. Corporate decision-making processes and delays due to people being unavailable or simply busy with more important things (to them) are often not taken into account. Getting an AI application to a ready state from nothing in just a few weeks is simply not realistic. Getting from scratch to production will take about one year most of the time. Give or take.

VALUE

The value that AI provides is often wildly exaggerated—this aspect will be discussed more in Chapter 10. You must consider not only the upside if AI gets it right but also the downside when AI gets it wrong—because it will. The quality of the process integration that allows your enterprise to act on the AI application matters significantly. The most important aspect is change management as your workforce must really use the application for it to offer any value at all. Claims of value must be examined closely with experts from the business who can speak to the realism of performing the required actions or decisions based on AI.

POSSIBILITY

The biggest claims in AI are impossibilities, at the current state-of-the-art. Innovation in AI comes from three main sources. Having more data allows the training of better models that are more general. Having more computational resources allows the training of larger models that are then better because they can capture more details. Refinements in algorithms allow us to train better models on the same data and with the same resources. The progress of the recent one or two decades has been accomplished mostly on the back of more data and resources. While some algorithmic innovation did take place, those innovations do not occur every few months but rather once every several years.

As far as generative AI, or large language models, are concerned, we have largely run out of data (the entire internet) and resources (hundreds of millions of dollars) for training. The incremental quality improvement from one version to the next is marginal and converging at the time of writing this book in 2025. What is needed is an algorithm-level innovation to make a step-change improvement forward. These innovations cannot be programmatically generated by force or funding. These are necessarily creative leaps of the imagination followed by considerable experimentation and fine-tuning.

Achieving human-like intelligence or the capability of reasoning either logically or causally is something that current models simply cannot do. Any claim that they can already do this or that they will be able to do it soon is simply incorrect and, frankly, irresponsible. The number of errors that large language models currently make is high enough that they cannot be used in many circumstances in which people would want to use them.

These applications are extremely impressive, but they are not intelligent in any normal meaning of the word. Most importantly, they may not be helpful or valuable to you and your enterprise in a real-life setting with all the non-ideal circumstances that reality provides. Other branches of AI such as computer vision and machine learning are considerably more mature than language models.

The Evolution of Hype Over Time

Things change over time and AI seems to evolve faster than most other fields. This is actually an illusion deliberately created by marketing. AI technology and capabilities have developed slowly over the years. What is developing over months and every few weeks is the commercial ecosystem of companies and product offerings—most of which are based on foundations that are, relatively speaking, much older.

Market research firms track the evolution of the field over time. Gartner even calls its regular tracking report "hype cycle," referring to the fact that all technologies are first overexaggerated and then overly vilified before they reach a stable level of operations.[1]

Such overview research is highly useful to keep the vocabulary, technology, and products in perspective as well as to track them over time. Among the general conclusions we may draw from decades of such research is that the speed of technological evolution has increased from the point of view of technology and AI companies. That is to say, if you plan to release AI products into the market, the pressure has increased. But if you are a customer of AI companies, the speed has not increased. The speed of adoption remains slow and occurs over a period of years.

A company adopting AI technologies a year or two later than another company in the same sector may get the benefits later, but it will not experience a dramatic outcome as a result. A company adopting AI earlier than others in its sector will see benefits earlier, but it will not be able to attain an unassailable advantage because of it. Exceptions, of course, always exist. These exceptions are often when AI is used to enable a new business model and the very way of doing business in the sector is changed. This has been done several times.

Popular examples include Uber revolutionizing the taxi industry or Airbnb revolutionizing the hotel industry. While the emphasis for both businesses is often placed on the simultaneous building of both the demand and supply communities, AI has had a major role to play in making this possible. Personalized suggestions on where to go, an accurate estimate of how far away the car is, detecting fraudulent behavior, and forecasting demand and supply are some core functions provided by AI. Absolutely core to both businesses is dynamic pricing—the automatic adjustment of prices based on the current state of supply and demand in relation to your inquiry.

KEY TAKEAWAYS

1 Create and advertise an AI community of practice. Sign up members, possibly in three groups: Interested users, leaders, and citizen data scientists.

2 Organize regular events and training courses to build a community, raise awareness of enterprise projects, and upskill everyone.

3 Consistently address the hype of AI and the fast-evolving nature of the startup ecosystem, putting all this into context for your enterprise.

Note

1   Gartner (n.d.) Gartner Hype Cycle. www.gartner.com/en/research/methodologies/gartner-hype-cycle (archived at https://perma.cc/LV8S-U9N7)

3

Identifying and Prioritizing Use Cases

Identifying Use Cases

"Individuals and organizations run into trouble because they too often solve the wrong problem. [...] The problem originally presented is rarely the most critical problem for the group to work on; oftentimes it is only a symptom, and a more urgent and important problem emerges as the group works on it."

MICHAEL MARQUARDT, *LEADING WITH QUESTIONS*

A use case is any specific task that you might want to accomplish with AI in your context. Transcribe a virtual meeting, summarize a document, forecast a price, detect a person on an image are examples of common use cases for AI. In any company, these typical use cases would be more specific still by putting them into context of, for instance, which price needs forecasting, for what purpose, with what accuracy, for how long a time into the future, and—most importantly—what would follow that forecast, which is typically a decision-making process. For a price forecast, the decision might be to buy or sell something earlier or later to profit from the price evolution over time.

Many consulting companies, keen to capitalize on the hype around AI, advocate running use case workshops in which teams around the company get together to think about what they could possibly apply AI to. These workshops tend to be great entertainment as people get to dream of the possibilities of a wonderful future in which many of the tasks they hate doing are miraculously automated. As the use case list must be assembled and presented, this is great business for consultants, with little accountability for anyone beyond making a great report.

The problems with these use case workshops, as with the broader discourse on the uses or value of AI, originate from multiple sources.

AI Is a Tool

The principal problem is that this use case workshop process puts the cart before the horse. In asking what you can apply AI to, you ignore the basic reality that AI is a tool. For example, let me propose that you invite a plumber to your house. Perhaps the bathroom water pressure could be increased for comfort, or the water flowrate lowered to save costs, or the total length of piping in your home lowered by 10 percent to save long-term maintenance costs. Beyond the prosaic, perhaps the plumber could build an entirely modern bathroom to impress your guests.

You get the point. You would never invite a plumber into your home unless you had a clear task that needed doing. Plumbing is a problem-driven issue. You get a hammer when you want to put in a nail, a screwdriver for a screw, and a wrench for a bolt. You know this.

Why do you know this? Because you've used these tools before and have some basic understanding of how home maintenance works, even if you are not proficient with these tools. AI is a Swiss army knife. It has several component tools in it that you can use for different purposes, but it is still a tool and its uses are finite. The lines grow blurred because most people involved in the decision-making process have never used AI themselves and so are not proficient in it.

While a plumber *can* build you a new bathroom, you may not *need* a new bathroom or be able to afford one. So, look for what you need to be done and not what you can apply your new shiny toy to.

AI Makes Me Feel Good

Asking people what AI could be used for often leads to use cases that offer emotional value rather than a business value. These are cases where people feel that a task is boring, onerous, or unpleasant and they want AI to automate it away. While perfectly legitimate as a desire, this does not mean this case is a good investment.

Let's look at an example. A maintenance measure on a machine is performed by a mechanic and a short report on the work done is produced. Based on certain criteria, a form may now have to be filled out and this is annoying for the mechanic. Having the form filled by AI would alleviate 15 minutes of drudgery. This is very desirable from the point of view of the mechanic.

As soon as you find out that this happens only twice a day for a team of ten, you can easily conclude that the overall time saved is quite small (300 minutes per day) and very likely not worth an investment in the technology while also accounting for the risks.

The mechanic has implicitly assumed that the AI system would "just work." This means that it would be able to fill the form on day one, out of the box, and that it would do a near-perfect job. Neither is true. This will involve work from the AI team and the mechanic as the domain expert.

Projects that would make certain people feel better come up frequently and they are often not good financial investments. And once they are conceived, they often have an unusually long life expectancy in the minds of those who thought of them, and thus become a liability for an AI team.

Focus on Time Saved

Many use cases that are likely to come up will have a time-saving motive. People naturally put themselves at the center of their world and think about how AI could help them personally. They have heard that AI is good at repetitive tasks and think about what they are doing often that AI could do instead. As in our prior example, these tasks usually do not occur often enough or to enough people to be commercially relevant to the enterprise as a whole. They are also often too nuanced for the AI to have a high enough accuracy to consider them truly "automated" without having a human checking every item.

If it all worked out though, what would happen if you saved someone 15 minutes in their day, twice a day? That person would feel good about this, that's for sure. The emotional value would be real. That may in fact translate to improved retention or workplace satisfaction. But, what it most likely will not accomplish is increased productivity. The time saved is just too short and too infrequent to translate into doing substantively more work.

If the AI were to save, say, two contiguous hours in a day, then you could argue that the person could genuinely do more work than before and so you may experience increased productivity. Cases of this nature, however, are far and few between.

In fact, one of the only widely documented cases that truly result in a significant productivity uplift is writing computer source code. LLMs assisting experienced computer programmers can as much as double productivity. However, the outcome is not necessarily as good. There is evidence that the size of the code is much larger and more convoluted than what a human would have designed. This makes it harder to understand, debug, maintain, and expand in the future. This therefore increases **technical debt** (see box) and so some of the productivity increases are, in fact, a type of loan.

Getting AI to help write source code is also known to introduce many cybersecurity vulnerabilities into the code that must be removed by competent humans.

CONCEPT: TECHNICAL DEBT

This concept refers to shortcuts taken during software development that led to someone having to come back later to "do it right." These shortcuts are common in proofs-of-concept where you are aiming to solve the typical case (not all cases) of a problem and usually not worried about security.

The debt must be repaid in the future. Due to the lapse of time, the future developer needs time to design the right approach and develop it considering all the code that has been built already. This is harder and more time consuming than doing it properly from the start. That fact is the source of thinking about this like taking a loan—there is an interest rate in effort that one must pay in the future or suffer vulnerabilities or inaccuracies.

Doing Something Nice But Useless

Use cases like transcribing meetings, summarizing them, and writing action items are very impressive and frequently cited as one of the premier use cases of modern large language models around the office. It's nice, to be sure. But is it actually useful?

Most meetings are transactional and the matters discussed ephemeral. If an action item results, it is often simple and clear. Keeping records of these meetings was typically not done in the past. Doing so now is possible but does not add real value.

Discussions often occur in a series, where meetings pick up on prior meetings. Note-taking systems usually do not take this into account and thus lack the context that the audience mentally has. The notes are there, and saved, but not saved in reference to a stream of prior notes. After dozens or hundreds of meetings have been transcribed, finding the information again becomes another challenging task for AI.

When action items are important, this may be great to have. However, the system suffers from known accuracy problems in that some actions may be on the AI-generated list that are not real action items (not too bad) or not all real action items may be covered in the list (much worse).

Knowing this, humans keep writing their own notes when it matters. Which raises the question: Why have AI take notes when it doesn't matter? Repeat this thought process for myriad other use cases of questionable usefulness. Language models can help rewrite a text you have drafted to improve the language or grammar, but you will need to check the changes to make sure the content is not changed. This will make the text better, but you will lose time and so reduce your "productivity." They can also help you write the text in the first place, but this is likely to contain factual errors that you will need to correct. To what extent this improves your productivity depends on the number of factual details in the text and the amount of available reference material. Translation from one language to another seems useful, but you may suffer from not being able to check the result at all without asking another person to do it for you.

Asking for the Impossible

It turns out that a lot of the ideas that readily occur to people are extremely difficult to deliver on practically. It's not always the science of AI that makes them hard. It may be that the data is not there, the skills are not available, the accuracy requirements are too high, or it would cost a prohibitive amount of money or time.

The hype behind AI has generated an expectation that AI is magic. It can deliver nearly anything, can do so automatically, instantaneously, and cheaply. None of this is true.

AI requires real work that is almost always a tight collaboration between domain experts and AI scientists. This work takes time and costs money. The work needs resources in terms of software and hardware that also cost money and may take further time in negotiating contracts, for example.

The really perfidious element, though, is accuracy. There is a common assumption that AI will get it right so very nearly always that once the system is ready, you can just let it run and witness the transformational impact on the enterprise.

Accuracy can be improved with more data, computational resources, and effort; that is true. However, the law of diminishing returns applies. You often see that you can get—and these are rough numbers—80 percent accuracy right out of the gate with very little effort. This can be increased to 85 percent or even 90 percent by doing some moderate work. Further increases to 92 percent are possible but require digging deeper. Getting to 95 percent might be a serious problem and need investment. Doing anything beyond that level is often not realistic for resource-constrained businesses unless the problem is a critical bottleneck with profound implications. The cost of accuracy increases exponentially.

Frequently, customers or business stakeholders ask for accuracy of 99 percent and more. AI experts must teach them that, generally speaking, this is unrealistic. It is achievable in some cases with a large investment in data acquisition, data cleaning, and AI science. Data acquired in the daily operations of a normal company is generally not big, clean, or representative enough to merit this. If an AI project's business case depends on the model being 99 percent accurate, this is a red flag in my experience.

An Alternative Exercise

In total, the exercise of asking for AI use cases will yield many use cases that, upon further reflection, will not be worked on by the AI team, nor would they offer business value for the enterprise as a whole. This does not only waste time, it wastes the political and emotional capital that has now been invested by all the people who took part. There is a latent expectation on the part of the participants that their use cases will be worked on.

A preferred approach would be to focus on the challenges faced in the various parts of the business. Ask what the problems are that constrain cost, revenue, quality, productivity, throughput, and the like. This process can be done in a series of one-on-one conversations so that people can open up and provide full information that they might hesitate to give in a broad forum.

Improving things with AI is good. It may not be easy since partial solutions will already be in place for known problems. In their solution, you may have to dismantle these partial solutions and interfere with popular "you have always done it like this" methods and tools.

The real value of AI is offered by doing things that previously were completely out of reach. You might dream by asking: If you had an infinite workforce at no cost, or an indefinite amount of time, what would you do? What is something that you would really like to do but could never do because it's completely out of range of resources? That's where AI can really help.

You can look at an image and detect a dangerous situation like a leaking pipe. You would have a hard time doing this at 20 frames per second, all day every day. It becomes prohibitive if you have thousands of cameras. With AI, you can realistically do this. The investment in a fleet of cameras streaming through an AI model in the cloud is realistic and often quite valuable.

You can listen to a customer complaint phone call and respond. It would be difficult to listen to all such calls for the last business day and totally impossible to listen to all "recorded for training purposes" calls of the past years. This data graveyard is suddenly no longer a graveyard at all. Using AI, this can be transcribed, analyzed, and summarized and statistical conclusions drawn that are actionable. The intelligence is human, to be sure, but the grunt work is done by AI.

Once you have a problem set, you can analyze it to see which problems can be addressed by AI and which problems are best solved by some other toolset. This will be a much better list of use cases and their business value is often more apparent. With the right expectations set in your conversations, this approach will not raise undue expectations that you're going to solve all the problems tomorrow.

Push or Pull

"The act of judgement that leads scientists to reject a previously accepted theory is always based upon more than a comparison of that theory with the world. The decision to reject one paradigm is always simultaneously the decision to accept another."

THOMAS S. KUHN, *THE STRUCTURE OF SCIENTIFIC REVOLUTIONS*

The opposite of running workshops asking what AI could possibly do is to think of the use cases at the central AI function and then tell the business units that they can expect to receive a widget that does this wonderful new thing. This is known as **push-mode**. The central function pushes its products out into the business. **Pull-mode** would be the central function asking the business what its challenges are and then solving those.

The push-mode generally results in failure and that is mainly due to three reasons:

1 *The problem never existed.* The central function usually lacks domain knowledge and deep insight into the problem or the processes surrounding the problem. They imagine that the problem is such and solve it. But the real problem is something else or depends on many more parameters than were imagined. Those extra data points may be inaccessible, non-existent, or expensive. This approach will yield wasted effort and cause the central function to lose credibility to rest of the business.

2   *The solution is unrealistic.* A real problem was solved but in a manner that the solution cannot be used. This is usually due to insufficient knowledge by the central function of the daily routine and business processes. It's either the software interfaces, user interfaces, or process interfaces that are hard to use or do not supply the requisite information that prevent the solution from working. Another aspect is that the real problem has several ad-hoc constraints that are hard to capture by AI and sabotage the whole effort.

3   *The business does not want to change.* Even if the central function delivers a realistically working solution to a real problem, the business may simply refuse because they were not involved, they never asked for it, they've always done it just so and it works for them, thank you very much! Often, they will now come up with a long list of reasons why this will not work.

In practice, it is hard to tell whether the solution is unrealistic or whether the business simply does not want to change. If the solution is unrealistic, you may have some hope to make it better. If really the business is simply refusing and you think that you can improve the solution, then you have run into the well-known sunk cost fallacy where you double down on an investment that has already failed.

Generally speaking, push-mode is bad. Both central functions and startup companies have failed when attempting to push their products into markets that simply did not want them or developed products that were not usable. The pull-mode is strongly recommended!

Pull-mode presupposes an open mind on the part of AI scientists and an investment of time and learning into each business unit. Chapter 6 will describe the pull-mode in detail and describe how to solicit challenges from the business. The key aspect is to be open-minded and learn from the business what its challenges are. Some can be solved by AI and those become your projects.

Clarity and Similarity Checks

"Be forthright in saying that the purpose of the conversation is to learn, not to judge."

"The primary difference between leaders and managers is that leaders are those who ask the right questions whereas managers are those tasked to answer those questions."

MICHAEL MARQUARDT, *LEADING WITH QUESTIONS*

At a high level, the pulling process starts with the AI team providing the business leaders with an explanation about why we are meeting to discuss their challenges—the reasons previously outlined. We are focused on identifying where costs could be saved, where revenues could be generated, or where processes could be made more effective by either removing or inserting new parts of the process.

A crucial aspect of this initial setting up of the discussion is to address the common fear that you are here to eliminate jobs. Generally, a person does so many different tasks during any one working day that it is very difficult to automate away an entire person. Perhaps one or two of those tasks can be automated and the people freed up to spend more time on the other tasks. It could be that, in aggregate, some people might be laid off in the process. Due to the point-solution nature of AI as well as its inherent nature of making mistakes, I do not believe this will replace large sections of the workforce as some pundits are forecasting.

In any case, saving a few salaries should not be the primary goal of an AI project. Far more value can be generated by looking at the revenue-generating capabilities or those parts of the process that can be created because they were inaccessible with purely human means.

Based on this introductory education and context setting in the workshop, let the ideas flow.

As challenges come up, be sure to clarify them so that the experts clearly delineate what they are. You are not looking for a detailed scope at this point and so do not go off on any tangents but get the experts to provide enough information so that the opportunity can be referred to in the future. You are also looking for enough clarity so that all the experts agree what it is and that it is important to them. Ideally the problem statement takes the form of one sentence that is clear and explicit. Here are a few examples:

1 AI will forecast the market price of a certain commodity, such as oil or wheat, for at least two weeks into the future to within $1 per pound to enable us to decide when to buy more.

2 Based on streaming cameras in our factories, AI will detect people entering red zones without appropriate safety gear and immediately alert the control room with an alarm.

3 Every morning, AI will supply an optimal route plan to each of our truck drivers on their mobile devices giving them a precise agenda for their deliveries.

These statements make it clear, at a high level, what the final outcome will be, both in terms of software and process. After reading this sentence, we have a vision of what the future might looks like, albeit a fuzzy vision.

Next, we check with the business experts for each of these challenges what the technical outcomes and financial outcomes could be. The primary things to look out for here are applications that depend on a highly accurate AI model and applications that are worth very little. Let's look at both of these in more detail.

Highly accurate AI models are difficult to make and maintain. By this, I loosely mean accuracies higher than 95 percent. They require lots of effort and time from AI and domain experts. They also require lots of high-quality data specific to the situation. As one of these elements may not be available, it is often infeasible to reach high accuracy. If the value of a use case depends on a model that is of high accuracy, it makes this a risky investment.

Most businesspeople have pet peeves. I do too. If AI could resolve this, that would be great for everyone. That in itself does not mean this opportunity is worth a lot to the company. As it is often very difficult to estimate business value, do not engage in a detailed valuation exercise at this point. Try to gauge the order of magnitude of the value here. Are we talking about thousands, tens of thousands, hundreds of thousands, millions? Over what time frame—one time or per year? If the benefit is per unit, how many units are there in the company?

The final question to ask the business experts is whether this application that they desire can be bought on the market or whether it has to be custom built. In general, buying is much faster and cheaper than building. If a company has built this application, they will have other customers who have already provided feedback and ironed out most of the bugs. You will be able to buy a working product right away as opposed to waiting for development and then working with the developers to debug it. This also frees up the internal team to work on challenges that cannot be bought.

Even if the business experts claim that no such application exists on the market, it is worth the AI team's time to invest a little research into the market to get a deeper insight. Often an application exists that is just not sufficiently well known by the business and can be leveraged. Another situation is that the application available on the market is not exactly what the business wants but does solve most of its problems. In those cases, it can be the best solution to pay the vendor to develop a custom add-on feature to close the gap.

Walking away from this workshop, the AI team internally must now think about the technical similarities between these projects. This conversation cannot be had with the business because the business usually lacks the technical AI or software engineering expertise to judge the overlap between their applications.

It is crucial to keep in mind that the vast majority of time, effort and money invested into an AI application is not related to the AI core of the application but rather to the data and process software interfaces and integrations. This is particularly true when we think of the full life cycle of an application beyond its initial development and onward into its maintenance over, possibly, many years and many users.

If two use cases can share a common data source integration or a common output layer to their users, this saves effort overall.

Frequently, one issue will be surfaced multiple times under different names and with different nuances. You may be able to serve several use cases by building out only one and adding a few small extras to it to cover the others. We will examine this idea in more detail in Chapter 4 on a project portfolio.

Prioritization

"Unfortunately, most organizations do two things very poorly—prioritize and focus. ... Organizations don't fail due to a lack of 'strategic' initiatives; organizations fail because they have too many."

BILL SCHMARZO, *THE ECONOMICS OF DATA, ANALYTICS, AND DIGITAL TRANSFORMATION*

Prioritizing use cases can be quite a headache because once a use case is on the list, it is important to someone. The question then becomes: How do you prioritize them and determine which ones to actually execute?

There are various metrics to consider when setting priorities. For instance, some projects might be prioritized based on how easy or difficult they are to implement, how long they have been on the to-do list, their perceived importance, or their urgency. Each of these dimensions presents its own challenges in terms of measurement and evaluation.

A rudimentary idea is to classify use cases by the amount of value they create. The three categories that result from this thinking are:

1 *Defensive*: If you do not do this, you will lose money. So doing it will prevent loss of money, market share, time or other resources.

2 *Offensive*: If you do this, you will benefit by some measurable amount that you can quantify reasonably well.

3 *Revolutionary*: Doing this, if it is successful, will create a qualitative change in the way your enterprise does business. You may not be able to quantify the benefit at this point.

From an investment point of view, it is probably a good idea to segment your portfolio of projects such that you have a good distribution between all three categories. The exact distribution depends on the quantitative value represented by defensive and offensive projects. Generally, it is advisable for the enterprise setting up a central AI program to split the portfolio into 40 percent defensive, 40 percent offensive, and 20 percent revolutionary projects where the percentages refer not to the count of projects but to the total investment of money made in these projects. As revolutionary projects tend to be more costly than defensive projects, the number of projects in the revolutionary category is therefore most likely small. While interesting, this division does not help in prioritizing projects.

The business cares about value and risk. They want a highly feasible large return on the investment in AI projects. It makes sense, therefore, to attempt to estimate both not in absolute amounts as this is difficult but in relative terms. Each dimension—feasibility and value—will be assessed by asking five qualitative questions, each of which will be answered on a numerical scale from 1 to 5 where 5 is the most favorable situation.

You assess the value of the use case by asking to what degree the business stakeholder agrees with these statements:

1 The use case is fully aligned with the enterprise business strategy. If the use case is fully aligned, you score a 5 and if the use case goes against the strategy, you score a 1.

2 The demand for this use case is high. If there is great demand, you score a 5 and if the AI team dreamed this up, you get a 1.

3 This use case strongly differentiates your enterprise from competition. If this is a clear differentiator that will be hard to copy, you score a 5 and if everyone else is already doing it, you get a 1.

4 Value will be realized quickly. For this, you have to define a realistic time frame in your industry and situation. If you can get a return early in that timescale, you get a 5 and if the opportunity is likely to deliver only its own costs in perpetuity, you get a 1.

5 Work on this use case will not require delaying or foregoing any other use cases. This addresses opportunity cost considering limited resources, other priorities, and dependencies. If this case can be assigned to a service provider or a new set of employees and so it does not consume any restricted resources, you score a 5 and if this use case will use up finite non-expanding resources internally, you get a 1.

Now you assess the feasibility with the next five statements in the same manner:

1 All capabilities needed for this use case are present internally at the enterprise. If you have all the necessary capabilities, both in terms of personnel and software/hardware assets, you score a 5. If you have none, you get a 1.

2 You are certain that you can get any missing capabilities in time. If you have already identified these capabilities and just need to sign contracts, you give it a 5. If you are unsure whether you can find them at all, you get a 1.

3 This use case is affordable. This addresses the resource constraint in budget and so is relative to the budgetary situation. If you can ask for more budget to get this use case done, it will not affect funding for other use cases and so it's not cost-prohibitive. If it eats into a limited budget and thus prevents other use cases from execution, it may be prohibitive. If doing this extends the budget beyond what you had, you score a 5 and if it takes up a substantial portion of a fixed budget, you get a 1.

4 Any enterprise risk of this use case can be mitigated successfully. These risks could include cybersecurity, intellectual property, privacy rights, and unreliability. If it is easy to mitigate all risks, you give it a 5. If it's difficult, you get a 1.

5 Any cultural risks of this use case can be mitigated successfully. These factors include fear of failure, bad incentives, silos, and user resistance. If there are no cultural barriers, you score a 5. If you anticipate significant resistance, you get a 1. Note that user resistance is by far the greatest root cause for AI project failure. If a use case scores low here, the use case should probably not be done as envisioned, no matter its other scores.

For each set of questions, you can now add the scores and divide by 5 to get an average score. The use case can now be plotted on a priority quadrant, as shown in Figure 3.1.

FIGURE 3.1 Priority quadrant

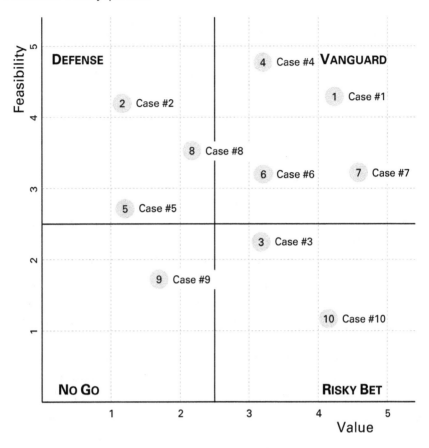

This plot allows a good visual comparison between different use cases. As is often done, such a chart can be divided into four quadrants. The low-value low-feasibility cases are a no-go territory as you do not want to do projects that are unlikely to succeed and drive little value even when they do. The low-value high-feasibility cases are the low-hanging fruits or the defensive projects referred to previously. The high-value low-feasibility cases are risky bets that might include the revolutionary cases discussed above. You should do some of these but not too many due to their risky nature. The high-value high-feasibility cases are the offensive cases referred to above and could be called the vanguard of what you want to focus on. These are lighthouse projects that you can realistically do and derive lots of value from.

Depending on some practical concerns such as budget and staffing constraints, you can now easily select those cases near the top-right of the chart and defend your choices visually to all stakeholders.

You can systematically evaluate and prioritize use cases by using this framework, ensuring you focus on those that offer the greatest value and alignment with your strategic goals. This structured approach helps us make informed decisions and effectively allocate your resources.

The assignment of scores for the ten questions is subjective, of course. It is also relative. In practice, I recommend that you assign these values centrally to the full portfolio of use cases. Whatever systematic errors you make, you will at least apply them consistently and so still arrive at a correct ranking. The attempt to get them assigned by the various customer groups in your company will require a lot of coordination and will yield widely different interpretations on what the scales mean.

To illustrate this line of questioning, a use case is briefly presented here. Suppose that you have a manufacturing company and the challenge you are faced with is the forecast of commodity prices such as the price of wheat, electricity, or gasoline.

1   The use case is fully aligned with the enterprise business strategy. The commodity itself might be totally essential to your company's core business, but this question is in relation to that forecasting model. Since you are going to buy it anyway and have limited storage capacity, the opportunity is aligned but not centrally important to the business strategy, unless the price swings are so great that buying it a little sooner or later makes a real difference. It scores about 3.

2   The demand for this use case is high. The factory workers are unlikely to organize a general strike to get this model, but the CFO's office might genuinely care about the financial benefits from buying at the right time. However, the factory planners are worried about inventory running low while waiting for a good price. It scores about 3.

3   This use case strongly differentiates your enterprise from competition. Supposing that this commodity is central to your product and you mainly compete on price, being able to buy cheaply at the right time would add a genuine differentiation, although probably not worthy of great fanfare. It scores about 4.

4   Value will be realized quickly. If this works, the positive return is virtually immediate since you will simply pay a lower price for the same thing. It scores 5.

5   Work on this use case will not require delaying or foregoing any other use cases. As forecasting this quantity is unlikely to involve any proprietary data or produce any highly secretive intellectual property, you could outsource this and so avoid any problems of a finite size team. As such, this probably has little impact on other projects and scores 5.

6   All capabilities needed for this use case are present internally at the enterprise. The commodity price is likely to depend on a variety of factors that you do not yet have data on, or even visibility about. Research will have to be done on how to transform data into a form that will offer a good forecast. It scores 2.

7   You are certain that you can get any missing capabilities in time. You will know about some external data. However, you are not quite sure about others and so this project poses a real scientific risk of perhaps not working out with sufficient accuracy or a sufficiently long forecasting period. The vendors you talked to are willing to help but have not done this before. It scores 1.

8   This use case is affordable. The project will cost effort to do research and purchase some data. It might take three months to prove whether it can be done at all. Whether this is an investment you are prepared to make depends on the possible upside and the available budget. Assuming that this is relatively affordable, it scores a 4.

9   Any enterprise risk of this use case can be mitigated successfully. This challenge is unlikely to pose any major cybersecurity or computational challenges other than the capability challenge that was covered in the prior question. It scores 5.

10 Any cultural risks of this use case can be mitigated successfully. The main risk is that the supply chain function will continue to order this commodity in the same way as before, thus preventing any value generation. Another risk is general trust in the forecast relative to inventory. Assuming that this has been carefully discussed and a healthy degree of willing suspension of disbelief has been produced, it scores 3.

On average, this challenge gets 4 of 5 for value and 3 of 5 for feasibility. Based on this assessment, this challenge might be accepted into the roadmap but not as a top priority due to the low feasibility. One would expect the portfolio to have opportunities with more confidence of success.

Developing a Roadmap

"Most companies have no clue why their customers are their customers. ... They don't know why their employees are their employees either. If the leader of the organization can't clearly articulate WHY the organization exists in terms beyond its products or services, then how does he expect the employees to know WHY to come to work?"

SIMON SINEK, *START WITH WHY*

Once the use cases and projects have been prioritized and put onto the roadmap, it is time to make some choices.

The first decision is to reject those projects that are either too difficult or not valuable enough, so that they are not worth pursuing. You must develop your own cut-off criteria for this. In addition to the roadmap, there are some other things to keep in mind.

A common way to assess difficulty is to determine if the basic functionality of the project can be proven in three months or less. This proof-of-concept (PoC) is a demonstration that a solution is possible. Generally, a PoC like this does not include input/output features, graphics, or nice-to-have elements but is restricted purely to the scientific heart of the problem. Often, it is based either on available products or prior work, often published in the scientific or conference literature. If you cannot prove the concept in under three months, this problem could be called a "research" problem and deemed too difficult for a commercial enterprise that is only interested in results but not interested in achieving research breakthroughs in the field of AI or related fields. Such tasks can then either be ignored or developed in partnership with research institutions by sponsoring Masters or Doctorate degrees.

A quick way to assess whether a project is valuable enough to consider doing it is whether the benefit is at least ten times the cost of execution. It is worth noting that it is likely that the cost estimate has been underestimated and that the benefit has been overestimated. This is simply human nature, and you will often be met with projects that cost double and more and deliver half and less. This conservative estimate will hopefully make sure that any project that meets this criterion at least turns a profit, even if it is small.

Projects should of course also have reasonable scores in most of the dimensions of the roadmap. If a project is totally misaligned with the company's strategic outlook, for example, you can probably safely skip it.

Every project that survives this basic shakeout process is now put on a roadmap. This is a fancy way of saying that you will get to it later. A roadmap can be as simple as a Gantt chart. The complexity can be given by three dimensions: The likely duration of a project in months, the number of people needed for the project, and the skills required of those people. You now have to consider that the number of people with those skills is necessarily limited, and you need to allocate projects to the timeline such that you obey these restrictions.

Crucially important is also the amount—in labor hours—of domain expertise your project will need from the side of the users or clients, and their actual availability. In practice, I have found it necessary that domain experts who are expected to contribute to the project need to be formally freed up, by their managers, from their day jobs by a measurable amount, such as 20 percent or one day per week. Only then will the project get the attention it needs.

Balancing these capacity constraints on top of the natural priority order in terms of business value, alignment, feasibility, and so on is not easy. This is especially true as many of these elements evolve as time goes by and the political boundary conditions change, and certain projects become more or less favorable or urgent. Taking into account the necessary communication, the maintenance of a roadmap for enterprise AI can easily become a full-time job for someone.

The roadmap is a zero-sum game as projects moving up will involve others that move down in priority and scheduling. The size of the team working on any one project is the variable leading to several projects being done simultaneously.

At this point, let's recall a popular middle school math problem. If it takes 5 builders 100 hours to build a house, how long will it take 50 builders? In real life, the answer is much larger than ten due to the unavoidable conflicts, coordination, organization, and communication going on. The same applies in reverse. How long will it take one builder? It will take longer than 500 hours because there are tasks that are hard for a single person to do by themselves, and perhaps they get stuck with a task in which they are not an expert. In short, there is a sweet spot for the team size. In software development, the sweet spot of a development team is around 5–7 people. You can shorten it to three or increase to nine. Smaller than three or larger than nine is not advisable as then you start to cut into some of these inefficiencies.

The ideal project team size places some natural bounds on the number of simultaneous projects that can reasonably and efficiently be worked on by a finite size staff. You should note that the ideal size of 5–7 people does not count the project manager, the product owners, the domain experts, or any persons mainly doing communication or organization. The number refers to the technical software and scientific developers. Accounting for the fact that any AI organization needs to have some of these other people on board, you can develop an easy rule of thumb that the number of concurrent projects is roughly one for every seven staff members.

KEY TAKEAWAYS

1   Don't look for use cases. Look for challenges and ask whether they can be solved by AI.

2   Get a clear description of these challenges and check their feasibility and business value to prioritize them.

3   Create a roadmap of projects over a timeline.

4

Creating Common Product Platforms and Organizational Programs

At the end of the road, you want to create lasting business value with the tool that is AI. To achieve this, users will be using a product that mostly or entirely takes the form of software with AI models and methods included. To make models and methods work in real life for real users, the product will also include a variety of other—non-AI—software methods and may include hardware components as well.

It is useful to distinguish between projects and products. Projects are team activities that create a product. Products are the result of projects. AI projects, like all projects, have a well-defined beginning and end. AI projects typically produce products that take the shape of software and thus must be maintained long after the project that created them is done. We will discuss projects in more detail in Chapters 6 and 7.

While AI scientists usually place a lot of emphasis on the project with its model training elements, the actual emphasis ought to lie with the product that is the value-creating engine for the business. That attitude is a win-win scenario, as this chapter will establish.

AI Product Pipelines

"How do you create value?—Is the most fundamental inquiry about strategy that anyone could ask."
"Be skeptical. You risk being drawn into practices that are not right for your company. Don't try to be great at everything."

PAUL LEINWAND, CESARE MAINARDI,

AND ART KLEINER, *STRATEGY THAT WORKS*

AI products have three main pipelines that move data through the product from end to end:

1 **Operations.** This is the regular pipeline that does whatever the product is meant to do. Data is ingested, transformed, interpreted with AI, and the output is presented to the user.

2 **Observation.** AI models are static once trained but the world produces live data that changes over time. This means models get worse with age. This pipeline continuously monitors the accuracy of the model and collects novel data points for later retraining. The observation is automatic and passive until the model accuracy dips below a threshold and then signals to the maintenance group that the model needs updating.

3 **Maintenance.** Once the observation process signals that the model is worse than you would like, a retraining and redeployment effort is started with the data that the observation pipeline collected in the meantime. This is usually a human-driven activity that can be partially automated, depending on the case.

All three of these pipelines contain multiple software steps. These include many of the normal software elements such as user authentication, cybersecurity, networking, data storage, compression, encryption, data formatting into a standard file format or database, and graphical display of the data or the AI results.

More intricate are elements such as master data management, ontologies, data retention strategies, and meta-data storage. Lastly, you must encode a certain business logic into the product that supports the decision-making process that typically results from the product.

The value of AI lies in automation. The value of automation lies in scale. Once made, an AI model and product can be deployed at a very low unit cost, and scaling AI applications is typically cheap. While this is true in principle, it is often not true in practice. AI products can be cheaply scaled up only when they have been built for scale from the beginning. This is mainly an issue of proper design of pipelines, which are all ordinary software and not usually themselves AI.

Assuming that your enterprise is pursuing multiple AI products, the pipelines involved will look very similar to each other. The basic software elements may, in fact, be virtually identical. The critical element is realizing and deciding to actually make them identical. In brief, the idea is to use exactly the same software systems and pipeline for a portfolio of AI products. This chapter will add lots of color to that idea.

Leaving aside the technical decisions and architectural considerations of such a design for the moment, let's consider the organizational ramifications.

Someone will need to define what these standard elements are and how to use and integrate them. This definition will have to happen at a technically precise level of detail, probably resulting in a documentation on both the application programming interface (API) and best practices standards of use.

The building blocks will then have to be made compulsory for the enterprise to use as defined to make sure that all groups working on AI projects and products actually leverage these common building blocks. From an enterprise perspective, common building blocks are great. They save money and effort in several ways. The enterprise will have fewer suppliers and license contracts to deal with. Enterprise licenses are often cheaper than many medium-tier licenses. Training on fewer products is easier. Maintaining fewer applications is more efficient. Common building blocks make the building of products faster and less error prone.

For project teams, however, they can restrict creativity and may lead to resistance. Teams will need to be included in the discussion of how these standards are achieved and why. Project teams need to understand that the enterprise is pursuing many projects and that their project fits into a greater vision and strategy. They also need to understand the similarities and differences between the projects so that they see the overall enterprise benefit of standard building blocks. Thinking about this, as well as long-term maintenance, will get most AI scientists and software developers bought into the basic idea.

Digging deeper, project teams have a lot of expertise and knowledge on the available building blocks on the markets. Very simple examples of such building blocks are file formats, storage locations, database systems, software libraries, programming languages, and web development frameworks. It is prudent to involve teams in the architectural decisions on which building blocks to choose. In making these choices, there are technical considerations, of course. Awareness will need to be raised with some teams that other considerations also play a role, such as licensing terms, availability of support, long-term software maintenance, an ecosystem of other customers who help to iron out bugs and collectively fund future development.

Together, we can agree on the common building blocks that we will all use.

Deployment Is the Key to Value

"Taking on a challenge is a lot like riding a horse, if you're comfortable while doing it, you're probably doing it wrong."

TED LASSO, TV SERIES

These building blocks need to allow the AI model to get the input data and put the result into the right place. This needs to happen so that we can make good models and certify that the problem has been solved. Eventually, the models need to be deployed so that real users can productively use the AI to accomplish a business purpose. Deployment is therefore the key to reaping business value from AI.

The key idea here is that model making and model deployment are very different processes which also often involve different people. These processes therefore require different building blocks. The scientific building blocks are used by AI scientists to prepare training data, select the right model types, perform feature engineering, tune the hyper-parameters, assess the model for accuracy, and gradually lead to a model that is good enough for its purpose. The deployment building blocks are used by the AI scientists and software engineers to drive live streaming data to the model to present the result to a user in some usable form while keeping everything secure, private, efficient, scalable, and robust.

If we were to make and deploy a single model only, the deployment part of it would take the large majority of the cost, time, and human effort. The point of this chapter is that an enterprise will have many models that it will want to deploy. Deploying each one separately would be a mistake.

Project Synergies

"Many of the numerous instances of entrepreneurial failure can be attributed to the fact that a would-be entrepreneur failed to consider the relevant conditions of interdependence between the component with which he happened to be preoccupied and the rest of the larger system."

NATHAN ROSENBERG, *INSIDE THE BLACK BOX*

Having raised the high-level ideas of building blocks and the necessity of deployment, you will need to look over your AI project portfolio to determine what kind of blocks you require. This task will need you to have a

FIGURE 4.1   Enterprise platform strategy at Oxy

User Interface/User Experience
Observability Platform

Enterprise Search and Chat	Services and Plugins	Structured Data	Control	Optimization	Computer Vision
Case #1	Case #1	Case #1	Case #1	Case #1	Case #1
Case #2	Case #2	Case #2	Case #2	Case #2	Case #2
Case #3	Case #3	Case #3	Case #3	Case #3	Case #3
Case #4	Case #4	Case #4	Case #4	Case #4	Case #4
Case #5	Case #5	Case #5	Case #5	Case #5	Case #5
Case #6	Case #6	Case #6	Case #6	Case #6	Case #6

Orchestration Platform
Data Lakehouse

good idea of the technical and business requirements of each project and so this design may take many weeks to complete. It is worth it.

To make this concrete, I will discuss the process that I went through in designing a platform strategy at Oxy, as seen in Figure 4.1. This will hopefully provide some ideas for how to adapt this process or design to your enterprise and its maturity status.

The next subsections will go over the diagram in Figure 4.1, starting at the bottom with the data lakehouse, going over the orchestration platform, deployment containers, the observability layer, to the user interface that finally provides the answer to users.

The Foundation of a Data Lakehouse

Data is the fuel for AI. This data must be stored somewhere accessible to us. These systems suffer from a lot of buzzwords that make it all quite confusing to most non-specialists. I will try to illustrate some main categories here for a basic understanding.

Data that we can conveniently structure in a large table, akin to a spreadsheet, is often called **structured data**. This often takes the form of numbers, dates, categories, and so on. The data may be stored in multiple tables to make better sense. An example is the record of my supermarket purchases. There is a table that lists all the products available in the supermarket. Then, there is a table of my transactions in which the items I purchased are listed by reference to the table of products. Various tables must be joined together to provide an output that would satisfy a person.

A system that stores such data is usually called a **database** or **data warehouse**. These are, by and large, the same systems. They are differentiated by their usage. Databases are typically optimized for real-time operations such as supermarket checkouts. Data warehouses are typically optimized for historical data analysis such as figuring out my long-term purchasing habits.

Data that is not structured is called, uncreatively, **unstructured data**. This includes a variety of data such as images, videos, audio, and text documents. This data is usually stored in its native format, which is to say that images might be stored as JPG or GIF files, videos as MPEG4, and text documents as ASCII text, PDF, or various other formats. The place where such files go is called a **data lake**. Just managing consistent and uniform file formats is a major task that may involve a multitude of format converters that may have to deal with a host of corner cases.

A data lake is a complex thing as these files need to be managed. Often, the meta-data for each file is stored in a database to get a handle on the entire collection. The files themselves might be stored by an **object storage** system that prioritizes scalability and managing unstructured data while databases are typically stored in **block storage** that prioritizes speed. Simply storing your data in **file storage** like on your personal computer is not efficient for an enterprise application. For unstructured enterprise AI applications, object storage is the right choice. There are multiple vendors of these systems, differing primarily in performance and various access features.

Putting together an enterprise system that consists of both structured and unstructured data then needs, of course, a **data lakehouse**, which is the combination of all these systems. This can come in two flavors. A **data mesh** architecture is a federated system in which different business units manage their own data products whereas **data fabric** is a centralized system.

Technologically there are plenty of intermediate flavors here. The point is that there are many choices to be made in how your data is stored and accessed. These choices are partly technological and they are partly organizational. For instance, the enterprise must decide how centralized the data ownership should be and who is accountable for what.

The datasets are typically documented in a **data catalog**, where anyone can see what datasets are available. The expectation is that datasets grow over time, are managed by someone, and have some data quality standards that are maintained by someone. Together, these three properties are the elements of a **data product**. On the open market, you could purchase or subscribe to a data product. Internally in the enterprise, this might be free of charge, but the expectations are the same. Every dataset intended for regular consumption must be maintained and this represents a cost.

In short, a data lakehouse is the foundation for an enterprise AI program. It is an essential prerequisite to be able to do AI at scale.

An Orchestration Platform to Make Models

Having high-quality data neatly in one place is the right preparation for making good AI models. From all this data, someone must now select the right part of the data for training and testing of a new AI model. The dataset may need to be enhanced by inserting human domain knowledge. That is done in two major ways. First, data can be labeled where humans say that this data point belongs into a certain category. Second, humans may transform data in a certain way based on their knowledge of the problem, known as feature engineering.

Examples of **data labeling** are characterizing all data between two timestamps as an indication of a machine breakdown or drawing an outline on an image to say that this is a car, a person, or a house. Examples of **feature engineering** are introducing a moving time window that takes the average over ten minutes of all the measurements or introducing a new variable that is the ratio of two measurements.

This enhanced dataset is then presented to a training algorithm for which a human has to specify what kind of model should be trained. For a neural network, for instance, the number of layers and the number of neurons per layer need to be specified. The training algorithm itself has a few parameters that must also be set. Traditionally, such things are done by writing computer programs in programming languages like Python.

Most domain experts do not speak Python and most AI scientists do not know the domain well enough to solve the problem themselves. To overcome this limitation, orchestration platforms were created. These offer a no-code interface to AI model training. That means that the human user does not have to speak Python or know many of the software or AI technical concepts to create a basic model. In many practical scenarios, these basic models are good enough for real-life use.

Establishing such a platform for the wider community of practice can significantly democratize AI across the enterprise and increase the number of people creating AI. Even professional AI scientists can get to a first version of the model very rapidly this way. The greatest benefit lies in maintenance—as no code was written, no code must be documented or maintained.

The orchestration platform goes on top of the data lakehouse and creates AI models. If necessary, these models can be improved through dedicated coding beyond a first version. Across a portfolio of projects and models, you can reasonably expect that you will only need to do so for one in three models, however.

Deployment Containers for Each Data and Output Type

Across an enterprise, the main data types you will encounter are:

1 *Numbers.* Many of them will be ordered in time and so are called **time-series**. Examples are temperatures and pressures measured by sensors in a building or manufacturing plant.

2 *Images.* Some will be standalone images while most will occur as frames in a video or streaming feed.

3 *Text.* Most text is likely to come in the form of documents and some will come as typed conversations with a chatbot.

4 *Audio.* Some audio could represent human speech while other audio could be based on various non-human noises such as machine sounds.

5 *Items.* This is not a traditional data type, but it occurs frequently in practical life. This is data related to an object (like a box being delivered to your home) or an event (such as you ordering that box).

AI, in general, detects patterns or continues patterns. Let's look at a few examples. For time-series data, you are usually interested in whether something strange occurred (detect an event) or in a forecast (continue the latest pattern). For images, you are interested in whether there is a certain object in the image or where it is (detect the pattern). For text, you are interested in finding a piece of information in the text (detect the pattern) or continuing the conversation (continue the pattern). For human speech, you want a transcription (detect the pattern) and for non-human audio you usually want to know whether something strange happened (detect the pattern). For item data, you often want to know what the pattern is so that you can plan better for it in the future, such as my purchasing behavior.

Based on this high-level analysis, most AI applications ultimately output one of four things.

First and foremost, they output nothing at all. Yes, you read correctly. The most common use case of AI is to detect certain undesirable events and since those occur rarely, the application outputs nothing most of the time. Popular examples are detecting whether workers are wearing their personal protective equipment, whether someone is breaking into your home, or whether a machine is about to break down. Vast amounts of data are analyzed 24/7 and only occasionally is there a meaningful output.

Second, they output an instruction to another machine. This might be an instruction to raise the temperature, to lock a door, to shut down a piece of equipment, or to start a piece of equipment. These applications communicate not with humans but other applications in their internal language, mostly an API.

Third, they output a notification. When something noteworthy happens, a human is notified of this pattern having been detected. This can be quite simple and take the form of a pop-up on a screen or a text message to a phone. It is then up to that human to take it from there and do something. This output may include highlighting the detected pattern, such as shading a selected area of an image to indicate the presence or absence of some object.

Fourth, they output the continued pattern in some meaningful form. Continuing a time-series could be a forecast of what will happen in the future, like a weather forecast. Continuing a piece of text is currently the most common use case using LLMs—you provide a prompt and the LLM continues the conversation by providing text of its own. Even though this looks like a conversation, deep in the model this is a forecast of the most likely next word in a sequence.

The reason for thinking about the data input types and output types is that many use cases in your enterprise will share these with each other. That means that they can also share the software components that process this type of data from the source, via the data lakehouse, and into the AI model. After AI is done processing the data, similar software components retrieve the answer and produce the desired output.

Suppose you want two computer vision applications. One detects whether field workers are wearing hard hats and the other detects cracks in pipes. Those two are the same use case. They sound very different but they both process video footage and both output a notification for a fairly rare event. In terms of software engineering, they differ only in the AI model itself. The entire rest of the processing pipeline of data is identical. That may even include the physical hardware such as the camera.

It is my recommendation that you look for these overlaps and design platforms for similar use cases that leverage that similarity and leave well-defined empty containers in the places of difference, such as the AI model itself. That way, the use cases can be brought from proof-of-concept into production very rapidly and cheaply once the platforms exist.

Comprehensive AI Model Observation

AI models are trained using training data. The model is therefore a distillation of that data. The live data that is presented to the model during its practical use is different from that training data. The whole point of AI is that we expect the model to behave well with novel data, but the novel data has to be quite similar to the training data for this to be true.

The critical question in practice is how similar the live data needs to be to the training data for the model to still be accurate. It is the testing data's purpose to determine that.

As time goes by, in most applications, the statistical nature of the live data changes. We often call this the **distribution** of the data. In training, the typical data point may have been around 5.0, and over time the typical data point in live data may start to be around 6.2. At some point, the statistical distribution of data points is sufficiently different from the training data to raise doubts about the model's accuracy. This does not need to be the average, but it could affect the variance or skewness of the data distribution. For images, the lighting of the object could be brighter or darker or the object could be closer or further away, and so on.

At those points, the model will need to be retrained using more current data that reflects the new reality. That leads to a sequence of models over time that need to be created and deployed.

Determining when this is necessary is called **observability**. There are dedicated software platforms for this purpose, and they attempt to determine the nature of the difference in the distribution to help AI scientists retrain the model to capture the new dynamic. They may also collect novel data points in between retrainings so that retraining can occur effectively.

This is an essential aspect of any AI deployment strategy as all AI models age over time and need regular replacement. The duration differs. Some models for consumer retail behavior must be retrained every week while industrial models may last for up to a year. That determination depends on your data sources.

Meet the Users with UI/UX

AI scientists operate in the shadows. All your users will ever see of your wonderful AI work is the final outcome as presented by the **user interface**

(**UI**). Depending on the application, that UI may be quite simple. It may, in fact, be a text message on their phone. It may be a pop-up window on their computer. It may be the sounding of an alarm siren on their work site. The UI is sometimes, but certainly not always, a complex software with graphics and diverse interactivities with the users. What the UI looks like for your users depends heavily on them and on the precise use case.

It is the responsibility of the AI experts to figure out, together with the users, what the UI must look like, feel like, work like to fulfill the use case and ultimately let the users leverage it to obtain the business value.

While interacting with the UI, the users have **user experience (UX)**. That experience, by its very nature, is emotional. Users do not experience business value, the reduction of costs, or the increase in revenues. To users, an application looks good or bad, it responds quickly or slowly, it provides all the necessary information for decision-making, or it does not, and so on. Most importantly, it is better or worse—in a loose emotional way—than what came before.

If the UX is good, users will use the product and thus leverage value. If the UX is poor, they will not. In real life, you will be somewhere in the middle. You may start towards the poor end and gradually get to the good end, but you must be careful to hold the users' patience and attention during the process of making things better.

Recall that there are no unhappy users. If your application's UX is sufficiently poor or the speed of improvement is too slow, your application, its use case, and its business value will vanish. Once it vanishes, it is hard to recover with a second attempt. Once users have experienced poor UX, that experience works like a scorched earth policy.

As the assessment of UX quality and improvement speed is an emotional assessment on the part of the users, it is critical that you stay in close contact with them to determine how things are going. Be aware in this emotional journey that the user population is not a uniform mass but that there are optimistic early adopters and pessimistic late adopters of your application, and everything in between.

AI Platforms: Buying or Building?

"Modern businesses ... depend on people voluntarily choosing to show up. ... They are paid for their efforts, but ultimately all employees are volunteers. ... Money does motivate people, but much less than culture."

PAUL J. ZAK, *TRUST FACTOR*

Having determined all the software components needed to make, deploy, and maintain AI models and their applications, it is time to decide whether to buy or build them. If you decide to buy them, you will need to decide on a vendor.

These decisions are not easy or fast. The landscape of offerings out there in the market is huge and constantly shifting. Most software vendors go through two to four versions of their software per year. Keeping track of features and roadmaps quickly becomes a practical impossibility. Let's look at some high-level pros and cons.

Buying Software

Buying has the advantage that you get working software immediately. The vendor of this software has other clients who have helped the vendor iron out the bugs so that you can assume the software to be somewhat mature. You can expect a degree of maintenance, training, support, and regular updates as well as emergency help should that ever be needed. In return, you pay a regular fee.

As you grow ever more dependent on this software, you gradually experience what can be described as **vendor lock-in,** which the vendor can use to increase fees that are not justified by a corresponding improvement in features. Vendor lock-in is primarily experienced when the **switching costs** from this vendor to another are comparable to the yearly fees paid to the vendor. You will feel "forced" to remain with a product that you feel is either too expensive or sub-standard compared to the market that has evolved in the meantime. It is therefore important to consider how you integrate products into your overall architecture. The more attention you pay to the interfaces between all the moving parts, the lower the switching costs will be, and your architecture can flexibly adapt to the evolution in the market.

Keeping switching costs low is the primary objective of good software architecture. Technical matters like latency and performance are less important. The other disadvantage of buying is that you only ever get an approximation of what you want. You must live with the product that is the outcome of what the whole market wants. The vendor is in control of the roadmap.

Building Software

Building has the advantage that you get precisely what you want. However, you must participate in the making of it, which takes a lot of effort over a long period of time. Many bugs will be found during this process. Since you are the only user, the bugs will be found and resolved at a slower pace than for a company that has many customers.

These activities will cost money that is often much larger than the yearly fees paid to a vendor who has a product ready to go. It is my observation that most services companies that offer to build products for you underestimate the building costs severely, often by an order of magnitude.

The reason for this is that the costs in most offer letters does not even consider some fairly major elements. All the costs of your domain experts providing advice, data, education, and review to the software engineers are not counted. The cost of providing and cleaning the source data is not included. Training and change management are not counted. The later iterations to get to a scalable product that the users actually like are (often intentionally) underestimated by the assumption that the users know exactly what they want at the outset and can articulate it, an assumption that is never true.

Hybrid Buy and Build

The software industry has reached a point where most tasks can be accomplished by a software you can buy. Building a complex system from scratch is no longer something that anyone will seriously consider. So, when people talk about building, they are referring to building a small aspect of a larger architecture or assembling components.

Generally, you will want to buy all the components that you are able to buy and only build what you cannot reasonably buy. You may also want to build the aspects that are your own special intellectual property that you do not want to reveal to product vendors who are likely to sell the insights to others.

Beyond this, what we will have to build is the whole workflow from the beginning to the end involving several bought software tools. This kind of building involves tying together the APIs of all the tools involved and adding business logic to them. This can be done in house or with a services vendor depending on your staffing situation.

Choosing Vendors

Having decided that you will need to buy some stuff, how do we choose the vendors? The first consideration is to carefully divide the market into product vendors and services vendors.

Product vendors sell software licenses. They have built a product, developed it further, and maintain it. They have multiple customers for this product. The fee is typically yearly dependent on some volume measurement of how much you use the product.

Services vendors rent staff. They may want to hide this fact by saying they want to pursue a partnership with you or develop so-called accelerators that are part products and that will lower the number of hours required to build what you need. They also have multiple customers but often this does not help you because the other customers are too different from you and the people working for you will not have been working for those others. The fee you pay is ultimately an hourly rate for the people you get, and the vendor is only as good as the people who end up working on your project. Getting a services vendor is loosely comparable to a recruitment process for a small team.

My recommendation for comparing product vendors is to first carefully define what you want the product to do and what is important to you. Design a test protocol to determine this. Only after having done this, look at the available products on the market. Do not design your test protocol based on the available features of the product you like best! Market analyst firms offer lists of product vendors for different categories of products that you can use as a starting point. A cursory examination of the features and a demo video will typically let you establish a list of a few contenders. It takes effort to really test a product and so you do not want to do this for more than two or three vendors.

Services vendors are very difficult to compare. They ultimately derive their staff from a similar labor pool and virtually all they do is pass along those people to work on your project. Even if you were to interview the individuals, you would find good and bad performers at every services vendor. The fees you pay per hour are comparable to people based in the same region of the world.

Differentiation between services vendors can be done based on four elements:

1 The company's technical leadership. The leaders of the vendor will perform quality control on what their team delivers to you. If they have good people vetting the content, that increases your chances of getting what you need.

2 Any accelerators the team has built previously. The existence of these partial products indicates that the company has internal knowledge and experience with similar problems, and these are software tools that you can test.

3 Domain knowledge of their staff acquired in past projects for customers like you. This domain knowledge is something you should be able to determine quickly in conversations with the individual developers about to be contracted to the team serving you.

4 References from similar customers that you should call to ask for their feedback. Some companies even specialize in a certain type of technology or industry, making them more qualified to work for you.

It is my recommendation to treat services vendors not as transactional providers but as longer-term partners. If they believe that the current contract is a transaction, they will naturally provide a second-tier team and not try their best. However, if they believe that you intend to foster a relationship that will provide a string of contracts for them over the longer term, they will provide a top-tier team and more quality control. As always, you get what you pay for, and you get treated well if you treat them well.

KEY TAKEAWAYS

1 Find the common software elements required to deploy your models at scale and design a holistic enterprise architecture.

2 Consider deployment over model building and design data pipelines and user interfaces such that you can swap models but keep applications running for years.

3 Buy what you can. Build what you must in cooperation with a curated ecosystem of vendors.

5

Managing Risk
and a Portfolio of Projects

Managing Risks in AI Projects

"Leading, however, means that others willingly follow you ... because they want to."
"When you compete against everyone else, no one wants to help you. But when you compete against yourself, everyone wants to help you."

SIMON SINEK, START WITH WHY

Risks in Communication

AI is ultimately delivered in the form of software and so many risks of AI projects are the same risks that any software project faces. Miscommunication is the primary risk in any of these projects. In particular, there are two forms of it that account for most of AI project failures.

First, there is a chance of misaligned expectations with the project stakeholders right from the start of the project. This risk will be addressed in detail in Chapter 6. In essence, an investment in time is required from all stakeholders in discussing and agreeing on a concrete and viscerally clear vision of the outcome at the end of the project.

Second, there may be a lack of communication with the project's end users throughout the project, leading to surprise at the end. This will be addressed in Chapters 7, 8, and 9. In essence, it is fruitful to include end users at every step along the way so that they are part of the process, aligned with the product, and have a hand in designing it so that it meets their needs.

Some of the nuances of these communications challenges are humorously displayed in Figure 5.1. Versions of this image have circulated since the beginning of the internet—it is unclear who originated it. This figure is my own version of the tale.

FIGURE 5.1 Some anecdotal challenges in communication for software and AI projects

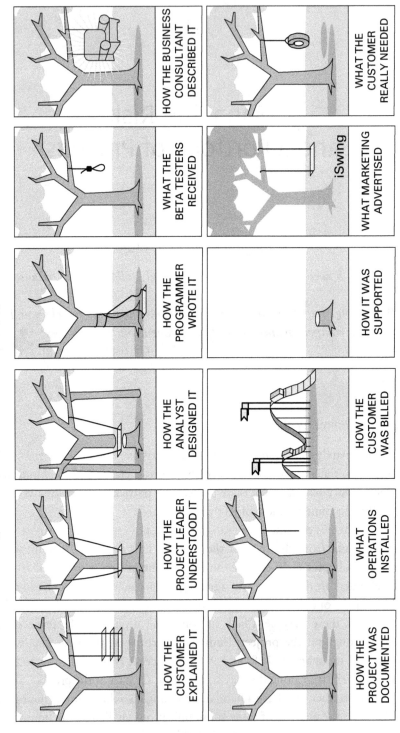

While the image attempts to be funny, it is perhaps rather tragic as these situations are quite common in software and, more recently, in AI projects. The individual tiles are self-explanatory for the most part, but a few elements deserve to be highlighted.

Consider first of all the top left tile. It is clear that this explanation by the customer—your business stakeholder—cannot be an accurate portrait of what they need. This is obvious in this image. In a conference room with a dozen people talking about technical details, this fact may be far from obvious. As an AI leader, you must listen carefully for incongruences and self-contradictions in what the stakeholders say to try to get as close as you can to the bottom right tile of what the customer needs. What the customer wants is almost never what the customer needs!

It is not your job to make the customer happy—by saying yes to what they want. Rather, it is your job to figure out what is needed to drive value and to lead the customer in a thought process of discovery that they eventually want what you know they need. In that process, you must avoid most of all your own misunderstanding (top row, second tile from the left) of what is needed. As the leader, if you misunderstand the vision, the project has already failed, even though you may not know it yet.

Most of the other challenges result from the leader misunderstanding what is needed and internal project management errors. Much of this is resolved by the agile method of software project management discussed in Chapter 7.

In resolving this key issue of creating a common and clear vision, you should spend more time describing the problem and then discussing what the solution should look like than discussing any technical details. For example, instead of focusing on the engineering details of the swing displayed in Figure 5.1, you should talk about the experience of the child using the swing and the parent pushing it. This approach can help avoid many misunderstandings and ensure you deliver what the customer actually needs. It may be helpful to go to the place where the problem occurs, for example to see the tree that needs a swing.

In some cases, it is helpful if the AI experts spend a little time attempting to do the tasks they are going to automate with AI to get direct experience of the problem. For example, I was once asked to create a predictive AI model for mechanical failures of a certain kind of pump (a rod or beam pump in an oilfield) but lacked any knowledge of it or the maintenance process around it. So, I spent a few interesting days with the maintenance crew driving around the field fixing pumps. In a short time, it became clear to me what the real challenges were and how the predictive maintenance AI model would have to be integrated into the larger workings of oilfield maintenance operations to improve the overall process.

Commonly, stakeholders want to start a proof-of-concept as quickly as possible with as few resources as possible. I propose that a slower, more thoughtful start has a significantly higher chance of success. Spending time defining the problem is an investment that will pay later in the project.

Risks in Data

Any analysis depends on the quality of the data available. Data quality will be discussed in greater detail in Chapter 11.

The risks to the project originate not so much from the data quality as such but rather from a lack of knowledge and understanding about what data is available and what its quality is. Early on in a project, stakeholders frequently make sweeping statements about their data, often assuming a certain status that may not be true. It is incumbent on the project team to check whether the data really is available. Frequently, you will find that the data is not in one but several systems and must be laboriously aligned and unified, delaying the project and generating extra costs. If the data spans several locations of your enterprise, you will often find that some locations have fewer sensors or a shorter history of data than others.

The monetary value of data has become clear only relatively recently with the advent of advanced analytics capabilities, including AI. Installing sensors in physical facilities, connecting them to IT systems, and recording their readings is a cost that was often saved during construction years ago. Retrofitting sensors now is greatly more expensive and may lead to lengthy discussions in a project. In case you encounter missing sensors, this sometimes leads to a prohibitive delay or expenditure in the project.

For industrial data, there is another version of the same tale. Sensor readings are stored in a database and occupy storage space on a hard drive. As sensors are read frequently, the readings are often very similar to the readings before and so the idea of a compression factor emerges—the new value will be stored on disk only if it is different from the last stored value by more than the compression factor. Due to the cost of data storage, many industrial facilities chose a relatively high compression factor. This factor now dominates the available accuracy of the readings. In some cases, it effectively erases many of the interesting fluctuations from which AI could have learned something.

Additionally, many facilities habitually store only the last few months of data in a rolling archive so that a longer history is simply deleted and no longer available.

One of the fastest and cheapest ways to reduce AI project risk is to examine your enterprise's choices with regard to data cadence and record keeping. As soon as possible in an AI project, look at what relevant data is actually available to you and in what form. The form may add time and cost to the project while the availability may go as far as making the project infeasible.

Risks in Science

In terms of the project life cycle, there are five technical and scientific steps to an AI project, bookended by two communication steps. Let's briefly go through these:

1 **Communication:** This initial step focuses on understanding the user, the problem, and the desired experience of the solution. It also involves determining the value of the solution.

2 **Data acquisition:** The fuel of AI is data. You need to find data that correctly characterizes the problem and is of good quality. This step may involve labeling the data, which can be particularly strenuous for images.

3 **Feature generation and selection:** This step involves generating more features through domain knowledge and selecting the most informative features. Features are channels of information, such as pressure transmitters or temperature sensors. You first generate more features and then select the ones that are most informative about the situation.

4 **Model selection and training:** Here, you choose the mathematical model for AI training, tune the hyperparameters of the algorithm, and perform the actual training. This is the main AI scientific step.

5 **Deployment and testing:** You test the model to ensure it is accurate and fast enough for deployment. You then deploy the model, observe its performance, and make necessary corrections.

Change management: This final communication step involves informing the user community about the new system, training them, observing successes and failures, and maintaining the system. Change management should begin at the start of the project but becomes crucial during deployment.

From a project manager's perspective, it's important to realize that 80 percent of the frustration occurs at the beginning due to the collision of different worlds, as discussed in the previous section. This frustration decreases as alignment is achieved. However, 80 percent of the project risk occurs at the end due to potential failures in change management and user adoption. Most of the human effort is required at the beginning for data quality and labeling, while most of the time delay occurs during the stages 3, 4, and 5 of the project, involving trial and error and waiting for computations.

The challenge is to keep everyone engaged through all these steps to ensure successful change management and user adoption. In my experience, the risk of an AI project failing during the scientific steps of the process—feature generation, model selection, and deployment—is low. If an AI project is going to fail, it is largely due to communication problems and sometimes due to data issues. The risks posed by these steps are not of project failure but rather of model accuracy and project duration and cost.

Feature engineering—the term combining feature generation and feature selection—is perhaps the most important technical step in any AI project. The raw data is virtually never maximally expressive about the situation or the underlying mechanism that produced the data, which is the entity that the AI is trying to model. This step is also necessarily collaborative between AI experts and domain experts since neither knows everything. The goal here is first to add features to the raw dataset (feature generation) and then to remove the irrelevant features from it (feature selection). This creates a new dataset that is streamlined specifically to the task at hand. So, what is a feature?

A **feature** is an element of data that is present in every data record. If the data is a tabular dataset, a feature is a column. For example, the data might record two pressure sensors to which the domain expert then adds a third column that is the difference between the two because the differential pressure is important for the physics of the process. For example, the three primary color components (red, green, blue) of a pixel in an image are features that could be combined into a single grayscale number if the task at hand does not require color information. In many cases, the features introduced can be very complex indeed and be little projects in their own right.

Beyond feature engineering, the work of model selection and hyperparameter tuning is largely a matter of trial and error that take time. One can invest rather a lot of time and computational resources into this process. While this will yield some degree of accuracy improvement, it is rarely a lot. You cannot normally expect more than a handful of percent of improvement here.

Model accuracy is often the most significant technical requirement in an AI project. Simply throwing raw data to the AI will produce a certain accuracy. Model selection and hyper-parameter tuning will provide a few more percentage points of accuracy. Beyond some amount of experimentation here, any further progress in accuracy comes at an exponentially increasing cost in time and computational resources. It is a myth to believe that every model can eventually reach a high accuracy. Due to the basis on incomplete imperfect real-world data, most AI models will remain roughly speaking in the 80 percent range and one should consider oneself lucky if the model gets into the 90 percent range.

The realistic step change in accuracy is provided by feature engineering. If you are uncertain about where to allocate a finite budget of time and resources in your project, allocate it to feature engineering and not to model tuning.

Finally, deployment is generally "just" work and so poses little risk of failure. However, this step is responsible for the majority of the project's total cost and effort! Deployment consists of creating all the software wrappings needed so that the AI model can be actively used by end users. As this will need live data to be made available and results displayed in a convenient manner, there may be substantive technical hurdles to be overcome. The primary risk to budget and timeline comes from incomplete or inaccurate scoping at the start of the project of the real situation of the IT systems involved. It may turn out that unifying data silos, writing data interfaces, or providing cloud communications turns out harder than anticipated.

Creating and Managing a Portfolio of Projects

"To the extent we have been successful, it is because we concentrated on identifying one-foot hurdles that we could step over rather than because we acquired any ability to clear seven-footers."

WARREN BUFFETT

The enterprise AI team will be conducting multiple projects simultaneously and many projects over the years. Chapter 4 already discussed the extreme benefit of identifying and using the various synergies between these projects. The goal is to have a modular platform architecture that allows reusable components to be used so that a new model can be packaged as a software application and deployed to users with little incremental effort. As mentioned before, the major part of the total cost of ownership for an AI application lies in the deployment stage—if that can effectively be removed due to existing platforms, you have supercharged your capabilities of problem solving.

Nonetheless, your capabilities are limited. Assuming a virtually unlimited number of potential AI projects in the enterprise for the foreseeable future, you are now faced with a two-dimensional allocation problem—one in time and one in personnel.

It is clearly infeasible to conduct every project simultaneously with a finite workforce. So, some projects will have to be worked on later in time, which raises the problem of prioritization and planning. While the prioritization challenge was addressed in Chapter 3, the ramifications of choosing to do a project *later in time* add some complexity.

You cannot allocate more personnel than you have, and you should also utilize everyone on your team at all times. This too is an allocation problem to be considered.

If the number of projects were small and nothing would change, this planning problem could be solved on a whiteboard after some discussion. However, for a larger number of projects, it gets harder. Additionally, requirements and effort estimates are frequently revised, leading to constant movement in the major factors of this planning process.

For every project, you will want to gather information about what skill-sets are required during the major milestones of the project for what amount of time. Together with the projects' prioritizations, you can now put each project on one overarching Gantt chart—see an example in Figure 5.2.

FIGURE 5.2 A time–personnel allocation Gantt chart

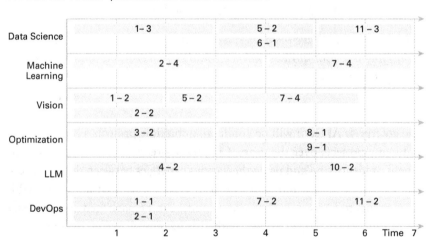

Time, from left to right, displays when the projects will be conducted. Vertically, every month, the personnel required adds up to the amount available in each of several skill categories.

Figure 5.2 has six lanes representing different skillsets from your staffing pool and shows projects over a seven-month period. Demand for staff is indicated by rectangles with two numbers in each. The first number indicates the project ID and the second indicates the number of people with that skillset required for this project and time frame. For example, project #1 requires three data scientists for three months, two vision engineers for two months, and one DevOps engineer for three months. Note that the number of staff in each skill category adds up to the same total in each month, except in the last month when the vision engineering team vanishes. This indicates a flaw in the plan because you have unallocated resources at that time.

As requirements change, the chart will reshuffle. If business unit stakeholders are unhappy about their position in the queue, you could display different scenarios if prioritization changed or the personnel situation changed. By accessing services vendors, you can get more people if the business makes the funding for them available. This situation is considered in detail in Chapter 14 and represents the major variable in this planning process.

There are multiple commercial software packages available that help in preparing and maintaining such a plan. While this may not be necessary early on in the lifetime of a central AI team, it will become necessary in the medium term as the number of projects grows. The need for planning will grow and, more importantly, the need to communicate and negotiate with the various lines of business with their competing priorities will grow as well.

For a large portfolio, a chart like Figure 5.2 will quickly become complex and therefore unsuitable for presentation to a business audience. This chart is primarily for internal use by the AI team for planning purposes. Another chart will be needed for the business.

The idea of a project portfolio is a crucial one to have and communicate. As the AI team goes through the process of finding and prioritizing use cases, the list of ideas will quickly grow. Each business unit or enterprise group will only be aware of its own use cases and will have the expectation that the central AI team will now proceed to work on them. It is at this time that you must make them aware of the growing portfolio of projects so that they can put their expectations in relation to everyone else's.

As will be discussed next in Chapter 6, it pays to carefully analyze each project to discover what it is about in detail before agreeing to work on it.

FIGURE 5.3 Overview of a portfolio plan in time for communication with business units

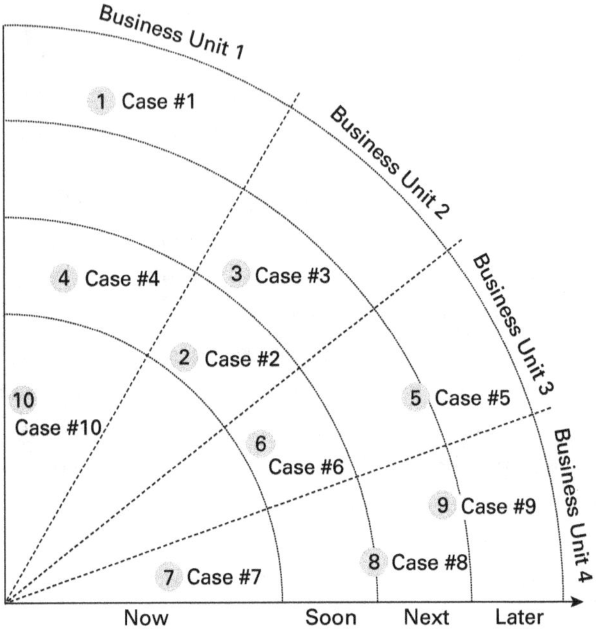

This assessment will also help to identify the major planning elements needed for the portfolio process—the time and people required, the costs and benefits to be achieved from the solution. With these elements identified—and, crucially, agreed upon—you can place the use case on the portfolio plan—see Figure 5.3. Finally, this plan can now be used to easily provide an overview of where the AI team is at in providing comprehensive value to the enterprise and assuring all stakeholders that they have not been forgotten but have their places on the roadmap of projects to come.

KEY TAKEAWAYS

1 The main risk in an AI project lies in communication. Pay attention to extreme clarity in defining the challenge and the expected outcome.

2 The most impactful lever is feature engineering, which is intensely collaborative between AI and domain experts, transforming both data and metrics into the right form.

3 While individual projects are important, the focus of the enterprise AI program must lie on the portfolio of projects that need planning in both time and personnel capacity.

PART TWO

Embedding the Enterprise AI Program into the Value Stream

6

Establishing a Project Charter and Implementing Design Thinking

Assemble the Stakeholders

"Most of the time, if managers take extra pains up front to discuss how they are going to communicate, many painful and costly faux pas can be avoided entirely. The problem comes when both parties proceed ... as if their style was normal and the other party was wrong."

ERIN MEYER, *THE CULTURE MAP*

The first serious activity for a new project is to identify all the stakeholders who will be involved. These are all the people who are needed to make the project both a scientific and commercial success as well as those who should not block its success.

The most important group of stakeholders are the targeted end users of the product that the project will develop. As this is potentially a large group of people, a few individuals should be selected. It is strategic to choose a few enthusiastic early adopters of the project's future outcome as well as a few critical late adopters in order to get a full spectrum of viewpoints.

The end users are crucial to the success of the project as they ultimately decide to use it or ignore it. As I've said before, there are no unhappy users. If someone is unhappy with the product, they will not use it. If someone is using it, they are at least somewhat happy with it. The only exception is when your organization can compel people to use a product against their will, but even then, they will resist, and this will cause other inefficiencies.

The next group are the business analysts who oversee the conversion of whatever the product will do into real tangible business value, such as more widgets, or lower resource consumption. They must be on board to deliver a business case for this project and bless the realism of its value proposition.

The third group are the domain experts or technical experts in the process that is about to be transformed. Your new product will either replace an existing process step or create a new one, but there is necessarily a process both before and after your new product. These people know that process and can provide the essential reality check to make sure that the product works in real life and in interaction with the rest of the process.

Naturally, AI projects need data scientists and AI engineers as well as IT staff to be a part of the project. In addition to all these people, you need their management involved at least up to a level capable of making budget decisions. Ideally, you would include management up to and including C-level leaders for their support.

In total, the group of stakeholders may easily be a few dozen individuals. If you end up with a list of 50–70 people, there is no cause for concern. However, if you end up with a list of less than ten, you have probably missed several key people and that will spell trouble later on.

Early on, psychology is more important than anything else. People want to feel good by being included, being heard, and being asked for their opinions and insights. You will increase their support greatly just by asking everyone's views. This takes time and effort, but it is not a waste because it is an investment in the future success of the project. It is also not a waste because you will learn many elements of the situation that you would otherwise not learn. Most people are passive and will not volunteer their insights until and unless you ask for them.

REAL-WORLD EXAMPLE

It is the early 2000s and a large multinational telecoms company in Europe is mailing tens of millions of paper bills to its customers worldwide every month. These are printed and enveloped in several printing centers in multiple countries. The process is slow, which leads to late sending of bills and thus impacts cashflow. A faster printing process is needed. Being innovative and listening to modern advice, a new spin-off company is created that can move fast as it is freed from all the corporate overhead. It recruits 30 staff members and gets to work. They design a completely new process and write lots of software. Two years later, the new system is introduced to great fanfare and greater hopes, overshadowed only by its monumental failure— the process is now several times slower than before so that one month's printing bleeds into the next, creating an ever-larger backlog of unprinted bills.

A "Ninja programmer" is sought after the spin-off company is dissolved and its staff fired. This lone individual does something radical and unheard of. He visits the printing centers, speaks to the line staff, and operates a printing machine himself for a few hours. This process takes only three days but determines where the problems truly lie that the spin-off's staff had never identified sitting in their shiny office. Within three months, this lone programmer writes a new printing software that solves these key problems and deploys it in one printing center, cutting the times in half.

In this, admittedly extreme, example, the approach of listening to the end users and developing empathy for them reduced the project staff from 30 to one, reduced the time from two years to three months, and made the difference between failure and success. The initial product was, of course, a minimum viable product that had to be improved. The full deployment software did take another year to make, but the core solution happened in months.

Running a Design Thinking Workshop

"A lot of times, people don't know what they want until you show it to them. Start with the customer experience ... Your job is to figure out what they're going to want before they do ... It's not the customer's job to know what they want."

STEVE JOBS

The process known as design thinking emerged in the 1960s and developed into multiple methods used for different industries. Specifically, to the software industry, the process became streamlined and brought into regular practice by Amazon Web Services (AWS) under the name "Working Backwards."

The root idea is to start with a vision of the final state of the product when everything is beautiful and the users are happy. This is developed together as a group and described in visceral, clear, and emotional terms so that all participants share a common vision and can articulate what it looks like. AWS likes to do this by writing what they call PR/FAQ, a press release/frequently asked questions document. This describes the problem, its solution, and provides quotes from the customer's leadership professing how the situation has improved through this solution. It then answers a few questions that end users or internal stakeholders are likely to ask and may even provide some wireframe visuals before the solution is built.

Two key ideas rule this approach. First, the end user is front and center of the entire process. Second, all stakeholders participate and create a common vision that they all agree with before the project starts.

In the past, many software projects were conducted only by technical teams working in isolation until a basic product was ready to be shown to users. This often failed as the team members often had insufficient knowledge of what users' daily lives looked like, what users needed, or what the real problems were. This type of process also significantly lowers the psychological willingness of the business and users to adopt the new product, even if it meets their needs. They feel as if they were left out of the process.

Having identified all the stakeholders, for our purposes the next step is to run a design thinking workshop. You gather them together into a room for at least one day, perhaps two depending on the complexity of the situation and the likely extent of disagreements. These workshops work well with between 20 and 40 participants. Having less than 20 is not ideal because the process will converge too soon and some important opinions are likely left out. More than 40 might become problematic due to logistics and aligning many people. I have successfully run workshops like this with up to 60 people.

As a group, you will walk through the five steps of design thinking. In each step, you split the group into subgroups of 4–7 people to discuss the question in detail. These subgroups then reconvene in the larger group to share their points and reach a consensus. The goal is to achieve universal agreement among all stakeholders, ensuring that every voice is heard and that you arrive at a written statement that everyone finds correct and complete.

Each step might take a couple of hours depending on the complexity of the situation and the number of stakeholders. This process relies on the active involvement of all participants. If there are lots of people present, this process might need multiple rooms for the breakout sessions. To capture ideas well, subgroups may use sticky notes on the walls to capture initial thoughts and then sift through them.

A best practice for generating ideas is silent brainstorming. This is when each person is asked to generate ideas by themselves and write them down on sticky notes. Only after everyone is done with this are the ideas discussed by the subgroup and then prioritized and condensed. This technique prevents more vocal people from dominating the group, drowning out quieter people's opinions.

The five steps in design thinking are as follows:

1 *Empathy*: You ask who the user is and what their pain point is. This involves describing the user, their role, their daily interactions, and the

friction or pain points they experience. You aim to understand what problems you need to solve.

2 *Define*: You define what the solution looks like from the user's point of view. This is not a technical specification but rather a description of how the solution should function for the user. It could include whether the solution works on a mobile phone or desktop, whether it has a typing or verbal interface, and how the user interacts with it.

3 *Ideate*: You break down the solution into concrete deliverables. This involves specifying the main technical milestones, such as the data required, interfaces for data import and export, and the computational elements needed for machine learning and AI.

4 *Prototype*: You define the minimum viable product (MVP). This is the smallest product that can solve the problem. The MVP should be achievable within three months and should be usable by a motivated user, even if it is not yet fully polished.

5 *Test*: You double-check that the MVP solves the problem. You revisit the initial user and pain point to ensure that the MVP addresses the core issue.

Even though we are talking about AI projects and making AI products for your customers and users, this process makes no direct reference to AI whatsoever. There is little discussion of models and computational infrastructure or whether the latest technique will offer some benefit.

The truth is that, by and large, users will not care at all if there is AI in the product or not. They care about having their problems solved. They care about how they can interact with the product. They care about whether all the information they need to make a good decision that they feel comfortable with is provided by the product so that this all supports the business.

This process is about the users, not about AI. It is about hearing everyone out so that you can get everyone on board that there will be a new product soon. It is about learning from everyone what the features are that this product needs. Most of these features will have little to do with AI. Users are primarily concerned with how the inputs make it into the tool and how the outputs are retrieved from the tool. They expect the tool to "just work." Whether that working takes AI or not is totally secondary.

In the workshop setting, the primary aim is to generate agreement on a common vision on these points. The IT and AI experts present need to provide a reality check on this evolving vision so that the group does not define some Hollywoodesque vision that cannot be realized, or an expansive vision that would take too long or too much budget. You want to end up with a common and realistic vision.

In that vein, you will also need to establish a high-level agreement in the group on the basic economics of this project. On the cost side, you will want the AI experts to give a rough estimate on how long this will take, how many people will need to work on it, and what items may have to be purchased for it. On the revenue side, the domain experts and business stakeholders will need to estimate what the potential benefits will be once that AI project is realized. Estimating revenue is often the hardest element of the workshop and also one of the most important as the business case will ultimately determine whether the project goes ahead.

Agreeing on a Project Canvas

"The greater danger for most of us lies not in setting our aim too high and falling short, but in setting our aim too low and achieving our mark."
MICHELANGELO

Having discussed everything with everyone at the workshop and achieved a common vision, it is time to write it down in a document that I call a **project canvas**. Every stakeholder is then required to formalize their agreement by signing this document. While this formalism may seem old-fashioned or quaint, there are some good reasons for it.

First, people have a habit of changing their minds. While this is expected and totally fine in the details, it is damaging if people change their minds about the major direction of the project or product. This is especially true if different stakeholders change their minds differently. This can lead to projects splitting into subprojects or lead to the scope of the project always changing just before it is done, leading to a never-ending project. This is often called **scope creep** or **moving target**.

Second, people either truly forget or strategically forget what they have discussed or agreed to.

Third, writing down what was discussed in design thinking workshops is helpful. However, you are all busy people and the number of documents moving across your desks and inboxes is probably very high. The chances that every stakeholder carefully reads the document, engages with it, flags their disagreements, and finally really agrees to its contents are vanishingly small. The simple step of asking every stakeholder to formally sign the document solves this problem—they will now read it and there will be a record of that.

The workshop itself will lay the foundation for the canvas, but it may easily take another two to three weeks of polishing to get everyone to agree to it. While that may seem ponderously slow and annoying, it is an investment that will pay high dividends once the MVP appears and the arguments begin.

When trying to write this document, it can be daunting knowing where to start and you may be tempted to write either a couple of lines or an entire essay. To avoid both, this book comes with a template for the canvas. It takes the form of three pages of template questions plus one page for signatures and another page for some explanations. You can obtain this template in PowerPoint format from the online resources that come with this book; the first page of it is displayed in Figure 6.1. This canvas is not a theoretical suggestion but something that I have tried and tested numerous times in my career and found to be helpful.

The sizes of the boxes in the template are intentional to prevent too much text being written. You are not trying to write software specifications or a statement of work here—you are trying to capture the common vision.

Let's go through the eight sections of the canvas and discuss what you are looking for.

Who Is the User and Stakeholder?

A vivid portrait of the user is what you are looking for here, not just a list of names or titles. You want to put yourselves in their shoes at a high level. Knowing their level of education, technical facility, work stress, location, desires, and typical personality will help guide us in designing software that will work for them.

Writing software with many options for users who have no knowledge of the choices will lead to failure. Providing an input interface that requires typing is useless for people driving cars most of the day.

To meet the users where they are at, you must know where that is. Describe it here.

The stakeholders of the project are individuals, and you are indeed looking for names here. These are the people who will later sign the canvas.

Why Are You Doing This?

The description of the user sets the context. The why is the real crux of the situation. Here you describe the pain the user feels that you want to resolve

FIGURE 6.1 The first page of the project canvas

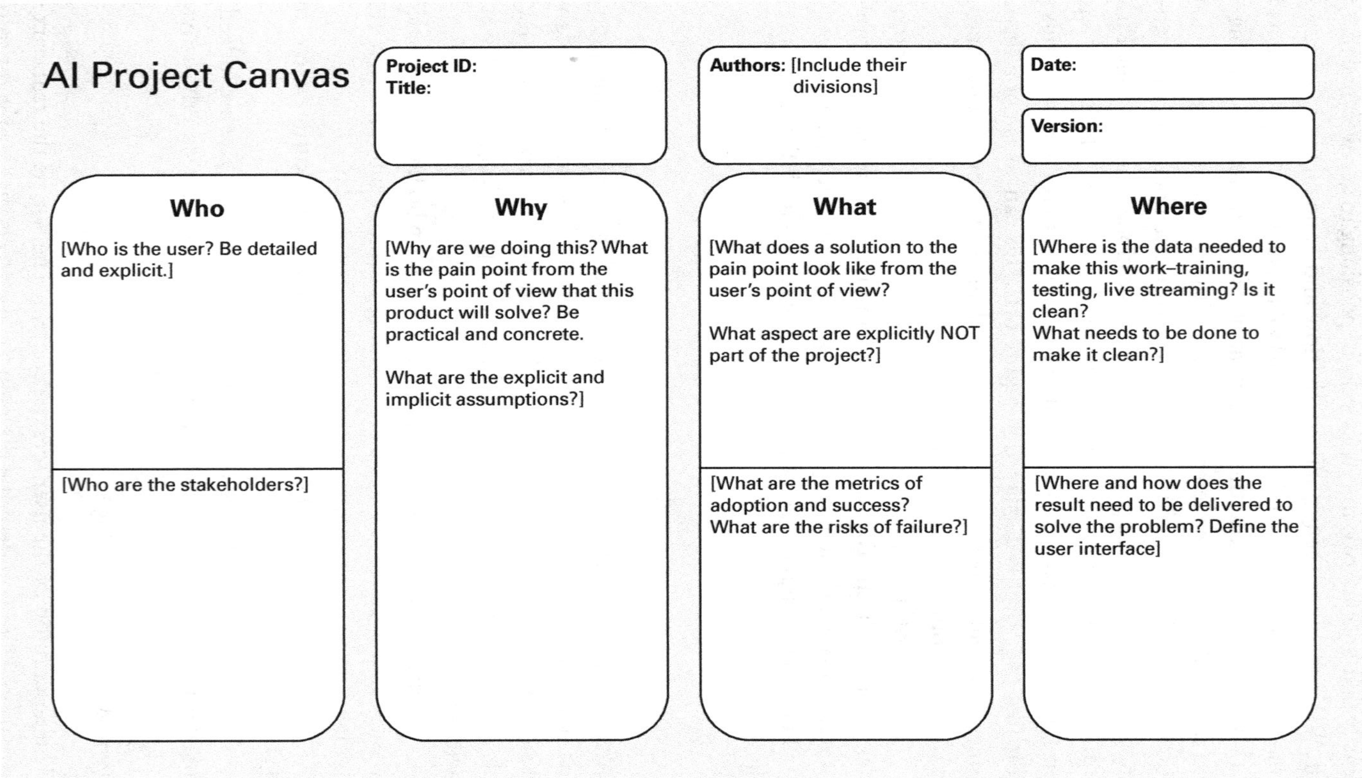

AI Project Canvas

Project ID: Title:	Authors: [Include their divisions]	Date:
		Version:

Who

[Who is the user? Be detailed and explicit.]

[Who are the stakeholders?]

Why

[Why are we doing this? What is the pain point from the user's point of view that this product will solve? Be practical and concrete.

What are the explicit and implicit assumptions?]

What

[What does a solution to the pain point look like from the user's point of view?

What aspect are explicitly NOT part of the project?]

[What are the metrics of adoption and success?
What are the risks of failure?]

Where

[Where is the data needed to make this work–training, testing, live streaming? Is it clean?
What needs to be done to make it clean?]

[Where and how does the result need to be delivered to solve the problem? Define the user interface] or alleviate in this project. Are you taking away some dangerous, annoying, repetitive, slow, tedious task or are you providing the user with some new technical superpower?

This section is looking for the pain point from the user's perspective. Not your perspective as AI experts. Nor the corporate perspective from a financial viewpoint—that will follow in the How Much section. You are looking at the users here. Recall that there are no unhappy users. You must make your users happy to have a chance at capturing business value because business value requires users to use the product. So, let's solve their pain point. Without writing an essay, you want to capture and bring out the essential features and causal mechanisms behind the pain point.

There are some common types of pain points. Some tasks or processes are slow. Users want greater speed to save time. Some tasks are so time-consuming that no one has done them. Users dream of having AI do it for them. This would open an entirely new product or service for the enterprise because it can now do something that was previously out of reach.

What Will the Product and Its Success Look Like?

Having nailed the pain point, it's now time to focus on the solution. What does the software that solves the pain point look like from the user's point of view? This perspective may likely not involve AI at all. It may merely be an input–output perspective relying on some core to make sure that the output is ok. It is not about technology but results. That is totally fine here. Your goal is not to be fancy or technologically advanced. Your aim is to solve real needs.

The primary focus of the user will be the input and output of the application. What information does the user have to provide and in what form does that happen? The more automated and easier it is possible to make this, the better. For every single input that you expect the user to provide, you should carefully analyze whether it is possible to somehow get that in an automated way from somewhere else. User experience degrades rapidly with the number of inputs that users need to provide and with the difficulty of providing them.

The output of the application needs to provide all the information that the user needs to take a decision or to do whatever job needs to be done. Some of this information may indeed be the output of your precious AI model while others may just be looked up somewhere. The manner of its presentation matters. Whether you display a table of numbers, sentences of text, or a colorful graphic, it must be adjusted to the needs of the user. Graphics are often the best way to provide complex information but need to be designed carefully to meet the user's needs.

Whatever the application does, there is a workflow or business process involving the user that starts before and ends after the application—the application only automates one step of whatever the business's entire process is. That creates two points of interface: An interface from the business process into the AI application and an interface from the AI application into the rest of the business process. These are very significant as they are interfaces to humans and systems. These interfaces are often not software but rather pieces of process.

As the new AI application either replaces or newly creates part of this entire process, the humans involved must adjust their daily work to this new element. Therefore, the software development team must recognize that to solve the problem, they must embed the new software in a process—not just develop new software. Embedding the software will include discussions, documents, training, and possibly various other elements such as the issuing of standing orders or standard operating procedures. This takes time, costs money, and poses a risk of user resistance.

In presenting the results of the AI to the user, the application must also consider any natural accuracy concerns that users have and should have. Merely outputting the result is often not responsible. Some indications of reliability are often helpful or necessary. This can take the form of probability or a confidence interval. Of course, such information must be provided in a way that can be understood by the real users of this application.

Users have accountability for their decisions and actions based on the application's output and so it needs to provide them with all the means needed to do so. This may go from simple record keeping to providing dashboards and analytics on the decisions taken, actions implemented, possible actions not implemented, and the respective real-world outcomes.

Providing contextual data generates trust in the AI application on the part of the users, which creates trust in their own decisions made on the application output's basis. In turn, that builds trust of the rest of the organization in this new human–AI process. In fact, trust is the greatest hurdle to and greatest ally of enterprise AI—without trust, the project will fail.

At this point, you are also asked to define the metrics that you will use in this project to measure the rate of adoption and the success of the project.

The product will only offer value if it is used. Being used is a flexible concept, however. Users may look at the product but not engage. They may engage but not use it for decision-making. They may make decisions based on the product but if they are all decisions not to do what is recommended, then the actions never result. Judging the usage is a tricky concept that must be based on the merits of the use case.

The metrics of adoption are related to the metrics of success, but they are not the same. Metrics of success measure how much value is generated by the product, relative to the situation without this product. Superficially, you might expect that the more a product is used, the higher its success. That is not true due to the accuracy of the AI models involved. If the product is fully used even when the model is mistaken, you also generate losses.

Depending on how you measure usage, mere usage may not result in actions that accomplish real-world success either. For example, if an AI model generates work orders for maintenance measures in a factory, the maintenance supervisor may accept these work orders. Here we have full usage of the product. However, if those AI work orders are not carried out because the foreman disagrees with them, no value is generated. Alternatively, if the AI model tells us to make certain changes in a manufacturing plant and all agree that this would make sense if it were not for some odd restriction that's in place in the world but that the AI model does not know about, then you achieve no success.

Finally, you are asked to list the main risks of failure. The main risks are rarely that the mathematics will not work out. Generally, the major general risks are of an AI project failing to deliver value. First, users refuse to use the product because it does not meet them where they are at. Second, users want to use it but are not able to because the product is too hard to use. Third, the product solves a problem that the users do not have. Fourth, some important input data is not available, or not available with sufficient frequency or quality. There could be many other risks depending on the specifics of the use case.

Where Are the Input and Output?

Every AI program needs input data and produces output data. In this section, you will ask how the input data is provided and how the output data must be delivered.

The input data is often provided either by manual human input or via some API from other software such as databases. This input data may be provided once in a while, or it may be streaming at a certain rate. A camera system may, for instance, provide images to the AI model at the rate of 20 frames per second. This places entirely different resource strains on your application than the occasional snapshot of a mobile phone camera.

The input data almost always must fulfill some stringent quality criteria. The most important thing is that the input data must be in the same range as the data on which the AI model was trained. If the live data on which the AI model operates is substantially different from the training data, the model will generally provide a poor output. Quality checks must be performed programmatically in the feeder pipelines to the AI model to ensure responsible processing of the data. You must be able to handle bad input data in some fault-tolerant manner—for instance by notifying the user that the input is damaged in some way. Those aspects must be built into an AI application for it to be ready for the real world.

The output may be provided to a human user in the form of a dashboard, a graphic, or a piece of text. This may be shipped to the user as a webpage, an email, an SMS text message, or in some other form. There may be a requirement that the delivery of the output must happen within some time window of receiving the input data—this is called the latency of the system.

Which Technical Elements Do You Need for Each Stage of the Project?

The technical elements that you will need the application to have are all the building blocks and connectors that you need to produce the final product. As you will almost never create an application from scratch but rather assemble it from previously formed parts, you will always have a list of proverbial LEGO blocks from which to build your system.

Typical examples of such elements would include software libraries like PyTorch, web programming libraries like AngularJS, databases, user authentication systems, and cloud processing components. These include several interfaces that either exist or must be built to make them talk to each other. Here, you figure out which building blocks to use and which of these must be built or customized by us.

The stages of the project referred to in the question are the four major stages of the project life cycle from proof-of-concept, MVP, proof-of-value, to production. The desire is to keep the number of elements for the first two stages to a minimum.

How Much Will It Cost and Benefit Us?

In building the business case, you now ask how much this project will cost us to build. This analysis will be provided by the AI team members as they look at the answers so far and analyze what needs to be bought and how much effort is needed to build what needs to be built. It is important here to take into account that the internal developers and the domain experts from the business are not free of charge.

As the project stages typically need very different technical elements and certainly very different amounts of labor effort, you will do well to estimate the cost up to a certain point. It is my preference to look at the cost up to and including the proof-of-value. The reason is that most projects make it to this point more or less according to plan within the reasonable assumptions made in the canvas document. At the proof-of-value, you encounter a true watershed moment that many projects fail. Even projects that move into production are often changed significantly in the meantime so that it is not responsible to judge the cost of productionization at this point in the life cycle, i.e., the very start.

Assuming an abundance of other proposed projects and a limited work-force, both of which are realistic in enterprise AI, any project will come with a significant opportunity cost. This cost should not be considered here since the canvas is an exercise about this project on its own merits and its opportunity cost is necessarily a feature not of the project but of the enterprise portfolio of projects that was considered in Chapter 5.

You should also ask how much the value of this product is, should it materialize according to the vision of the canvas at this point. This must be provided by the business and is usually the most difficult discussion of the entire workshop. In absence of a concrete use case, it is difficult to provide any specific guidelines for calculating product value. I recommend preparing the stakeholder audience prior to the workshop by priming them with thoughts about the business value of the final outcome. More on that in Chapter 10.

Often, you find that costs are underestimated and value is overestimated. That is why it is prudent to require that a product be worth at least ten times the cost of producing it so that after the true cost and benefit emerge, the product is at least at the breakeven point.

When Should You Do What?

Next, you think of a high-level timeline of major activities in the project. Apart from the four stages of the project life cycle mentioned previously, what other major activities need to happen in your project? Perhaps a field trip to a work site or a factory is required, or people need to be trained on some product. It could be that a product needs to be purchased and getting procurement to agree is on the critical path of the project. Some organizations might be required to approve some action or data access.

What Will People Think?

Finally, you imagine for a moment that the beautiful future with your new product is a reality and that all that value is being leveraged. How will the users feel about this? Let us quote these users when they express their experience at this point. You are trying to capture an emotional response about the improvement in their work lives from having this new tool. They will not be motivated much by the business value leveraged. They will feel good and happy because some boring tasks are now automated, some dreaded actions no longer need to be done, information is now at their fingertips and quickly, they can do something much more enjoyable while the computer does what they were doing before, and so on.

AI products have financial value to businesses and emotional value to users. If they do not have either one, they are not viable. Users will not use a product unless they feel better with the tool than without it. Emotional value to users turns into financial value to the business. Not the other way around.

This final step is usually fun for the workshop participants and so you end the workshop on a high note with people imagining the great experience that they will have in the future. That is the essence of generating a clear and shared vision of the product that you will now develop together.

Have Stakeholders Sign the Canvas

Congratulations on filling out the canvas. It's hard work to get this far. One more important step remains and that is getting all the stakeholders to—formally—agree. A virtual electronic signature service is great for getting all stakeholders to sign the canvas. This simple step accomplishes several important objectives. First, all stakeholders will definitely read the canvas if they are expected to sign it. Second, if anything is wrong with the text, they will flag it. Third, everyone is now on record as giving their formal approval of the vision.

Months later, if people change their mind, you can point them to the signed canvas to bring them back to the original vision. That is also the reason that the canvas is written at a high level. We want to constantly modify the details in agile project management, but we do not want to change the overall vision and problem statement at all during the project.

Even though it seems quaint, old-fashioned, bureaucratic, and overly burdensome, I recommend this strongly as it streamlines the process and prevents alignment problems later on.

To Pursue or Not to Pursue?

"To the extent, as significant as it is incomplete, that two scientific schools disagree about what is a problem and what a solution, they will inevitably talk through each other when debating the relative merits of their respective paradigms. In the partially circular argument that regularly results, each paradigm will be shown to satisfy more or less the criteria that it dictates for itself and to fall short of a few of those dictated by its opponent."

THOMAS S. KUHN, *THE STRUCTURE OF SCIENTIFIC REVOLUTIONS*

Setting expectations is one of the most important parts in the initial discussions. One crucial expectation is whether this project that is being canvassed will be implemented or not. Be clear to all stakeholders that this is uncertain until the canvas is finished. It is irresponsible to decide on the fate of a project while we do not yet know what that project is, how much it will cost, how long it will (PoC) take, and how much value it provides. Aligning all stakeholders on the answers to these questions is the point and purpose of the canvas process.

With these answers in hand, we now ask whether to pursue the project or not. This decision cannot be taken purely on the merits of the project itself. That is because resources are always limited and there are other projects clamoring for attention. The decision to pursue a project is therefore a decision at the portfolio level, which we discussed in Chapter 5.

If the project is pursued, there are generally no problems at this point. If, however, the decision is taken not to pursue this project, it is important to justify this decision carefully to prevent hard feelings and to prevent rogue projects from sprouting up.

Rogue projects are problematic principally because the full life cycle costs of AI projects are often radically underestimated. The initial development costs, up to and including the proof-of-value, are clear. Productionization is often more difficult and resource consuming than estimated. Change management is typically ignored entirely from a budgetary perspective. Finally, both model and software maintenance over the long term are not considered. These considerations often lead to significant expenditure (as we cannot call them investments) on projects that never succeed in delivering value.

KEY TAKEAWAYS

1 For each project, carefully identify and assemble all relevant stakeholders and hold a design thinking workshop.

2 Document the workshop in a written document and have all stakeholders formally approve it.

3 Decide whether to pursue the project on the merits of its value relative to its cost and risk to make and maintain.

7

Project Management and Agile Scrum

Project Stages and Milestones

"No one buys AI or technology. They buy an outcome or an experience."

VIN VASHISHTA, *FROM DATA TO PROFIT*

After defining the problem and drawing up the project canvas, the next question is often about the timeline for availability and the mechanics of getting the job done. The answer depends heavily on the type of IT project required. Let's first clarify some terminology and the key stages relevant to project timelines. Your projects will run through four major life cycle stages.

Proof-of-Concept: It Is Possible

The **proof-of-concept** determines whether the solution to the problem is possible. It is a scientific question, asking whether AI or ML can model the scenario, whether the data is expressive enough, and whether you have the technology to do it economically. In many cases, scientific viability may be obvious because this project has been done before, perhaps by others, and so a true science exercise may not be necessary. In most cases, the PoC degenerates into two other versions. It is a best practice to be very clear on what a project step is meant to accomplish.

The first version is the **proof-of-capability**, which is a practical demonstration that a vendor's product has all the features and functionalities required to do this project. Note that this is not a science exercise but rather a product demonstration or test. As such, this can be done far faster than a true PoC.

A **proof-of-competency** is the demonstration that a service vendor's staff has the required knowledge and skills to conduct the project. This is also not a science project and may turn into a modified form of a job interview to determine the suitability of the people involved. This can be accomplished rapidly if done correctly.

Expediency recommends that the project sponsor be very explicit about what is to be proven in what is often called a PoC. The specific proof point determines not only what people do but also how long it takes and the terms of a contract needed to get it done. When dealing with vendors, it is a good idea to let them know whether you are concerned with the quality of your own data, with their product features, or with the quality of their staff. The result of the PoC critically depends on that. Uncertainty on this is a frequent cause of a project taking far longer and being ill directed towards a decision-making point.

In terms of timeline, if you are proving a scientific concept, you want this to happen within three months. This is a rule of thumb that is now well established in the industry and has been shown to be a watershed criterion. Proofs that take longer than this generally fall into one of three categories:

- The concept is so difficult that it cannot be proven in three months and so the proof is really a research project that may take a much longer period. These projects may not be suitable for a commercial project in the first place and may need to be assigned to a research-and-development (R&D) lab, or to a university collaboration.

- The concept is ill defined so that proving it is not possible. This may be due to ambiguities in which the team is uncertain what the outcome is, or the concept may be a moving target that changes whenever proof gets near.

- The foundation is not ready. If the data or the teams are not ready for the project, the concept itself cannot be proven. The data may need significant and time-consuming cleaning, formatting, transportation, unification from silos, and so on to enable the proving to happen. Teams need to be supportive and available to work on the proof project.

It may seem silly to assign a specific period of time to a PoC. What if my project is a bit harder? The variable in all this is the team working on it. If the problem is harder, use more people, but stick to the three-month rule.

If you are in a capability proof, this can typically be done within three weeks by preparing some test data and carefully agreeing on a testing procedure. If competency is in question, this can typically be done in one or two weeks by testing the staff in a systematic manner.

Is it possible? If yes, you go on. If no, you discontinue the project at this point. From a portfolio perspective, you would like approximately one-third of your projects to fail this stage. If all projects pass the PoC, you are not innovative or aggressive enough. Should you find yourself in this situation, you ought to reexamine the process by which you accept new projects from the canvas stage (Chapter 6) and adjust the acceptable level of scientific risk, or, if this is a vendor's product assessment, product risk.

Minimum Viable Product (MVP): It Works

The MVP is the PoC plus enough software interface for a user to determine in a real-life scenario whether the solution works. At this point, the solution is still very basic, not user-friendly, and lacks polished graphics, but it demonstrates that the job can be done.

Typically, you need to be able to feed live data into the scientific core of the application and have a human-readable output. Those two interfaces turn the PoC into an MVP.

MVP is a **product** in the sense that it is a usable software that runs. A user can use it under normal circumstances. The MVP is **viable** in that it will run well enough to solve the core of the problem. It may not do so quickly or comfortably, but it does work. The MVP is **minimum** in that it delivers just enough functionality to address the problem.

Defining the features of an MVP is a careful exercise in identifying the barest bones core of the problem and the minimum amount of integration and interfacing required to be able to run the PoC with a real user in real-life conditions. The exercise will involve interviewing several of the target users to determine what their everyday job functions look like and how they will expect to interact with the software.

The timeline for an MVP is a further three months after the PoC. The circumstances are the same as before. If it takes longer than three months, the features of the MVP are not minimum and you should reduce them. The variable is also the number of developers working on the MVP to be able to conduct more work in the fixed time frame.

The purpose of the MVP is to be able to conduct the third stage in this process, the proof-of-value.

Proof-of-Value: It Is Valuable

The PoV is an activity, not a product. Users take the MVP and use it in real-life scenarios, not constructed academic ones. There might be only one or two users and one or two sample cases, but the goal is to observe and quantify the actual value generated by the MVP. This helps reassess whether the value is good enough and meets expectations.

Ideally the users would have been involved in the process of creating the MVP so that they are engaged, involved, and motivated to use it to prove its value. Involving them early also makes sure that they know what they are doing and can use the MVP—remember that it still is clunky to use as it is minimum.

Together with business stakeholders, these users would have planned out their testing while the MVP was being made so that they can seamlessly start with a testing program. As such, the PoV should take one or two months to complete.

At the end of the PoV, you would like to have a concrete numerical assessment of the value generated by the MVP in the selected cases. This might be in terms of additional revenue earned, lowered costs, time saved, risks lowered, or some other tangible result. As this might be complex and, often, controversial, the measurement methodology should also be discussed during the creation of the MVP, so that once the PoV begins, all stakeholders are clear on how value will be assessed.

The last step addresses the scale of the value. No matter how well the MVP does in the PoV, the value will be small as you are testing the application in carefully isolated cases. What you must do now is to estimate the full value if the solution were to be scaled up to full and permanent use by the whole organization. You should also think about how to scale up the value in advance, as this can be controversial too.

For example, imagine you determine that the value of damage detection for a certain gas turbine is $100,000, if the damage is in fact correctly detected. Now you must estimate the number of times that this occurs for which you can use past repair records. You know how many machines you have. Next, you determine whether the $100,000 is a true average or if that instance is more of a cheap or expensive case. Lastly, you must consider the differing costs of false-positives and false-negatives. Some of these choices are not easy and some statistics may take time to look up.

At this point in the process, you will have taken 7–8 months at most to put this together and you will have spent approximately 20 percent of the total cost of the solution. The other 80 percent will be spent in the last step of the process, in productionization. This stage gate is therefore very important.

The value must be carefully considered in relation to the cost of creating the full solution alongside the cost of maintaining the solution over the long term and the willingness of the users to use the solution in their everyday work. Failure to get user adoption is the principal cause of death of AI products, as you have observed before.

The decision is now an investment decision to invest the rest of the funds in the product in return for the value plus or minus the risks—the main risks being attaining user adoption, having correct estimates of cost and value, and being able to deliver a sufficiently correct solution. The last point is on your list of risks because the AI model accuracy is often, at this stage, not yet the highest accuracy attainable and considerable responsibility is often placed on the AI team and their ability to raise that accuracy during production.

Another aspect in the investment decision is the opportunity cost. You will be faced with limited staff and budget so that pursuing this project prevents you from pursuing another. Within the general topic of AI at the enterprise level of a mid- to large-sized company, the opportunity cost is a real concern as there will certainly be more opportunities than resources will allow you to pursue—probably even if one considers several future years of portfolio development.

You should expect roughly one quarter of all uses cases that make it to this point not to be promoted into production. As before, if all use cases are promoted, you are not aggressive enough in your portfolio and are not pursuing a diverse enough enterprise portfolio of projects.

In fact, the principal point of going through this staged process is to fail projects out! If every project were going to be fully developed into production, you might as well start full development projects from the start. As this is expensive and error prone, you use this process to weed out the ones that are technically and fiscally challenging.

Production: Let Us Do This

Next comes production. Here, you take the clunky MVP and make it a real product that is user-friendly, seamlessly integrated into workflows, and equipped with polished graphics. Building from the MVP to production requires significant investment. This is the most expensive stage in terms of effort, time, and money. On average, this stage costs four times as much as the prior three stages combined. If planned well, many aspects of productionization can be done in parallel, so that this effort does not have to take four times the duration of the prior steps.

Moving from MVP to production, you must consider three main collections of features.

First, you have the features of scalability. To use the solution with many users over many cases, the solution must include security, and stability features such as user authentication, encryption, data access privileges, cybersecurity, stable and secure data pipelines that can recover from a variety of issues, and so on. The focus here is having software that runs and can ingest its data.

Second, you have features of user interaction. The software must be integrated into the larger workflow of the users so that the work before and after the AI product feeds into and out of the solution in a seamless way. If users must work hard to feed the product or to extract answers from the product, they will not use it. If users make decisions on the product's basis, the product must provide not only the AI answer but sufficient context to support a good decision-making process.

Third, you have features of accuracy. Most users expect software to be accurate. The fact that AI models are necessarily less than 100 percent accurate is culturally difficult. Models making mistakes is not a "bug" but an essential and unavoidable element of using AI. Users must know and be somewhat comfortable with this situation. The solution must support them in this by providing helpful information such as confidence intervals, probabilities, and other information that allows the user to know what they are dealing with.

In the cases that AI does not produce the perfect or expected result, it is desirable that the answer be very close to the real answer in some meaningful way or that the AI system recognizes that the answer is wrong and can simply alert the user. The combination of these two modes is called **graceful degradation**. If an AI system outputs a significantly wrong output without giving the user an indication that it is wrong, it is understandably not graceful. It is not always possible to design for graceful degradation, but it is a best practice to work hard at creating it wherever possible.

There will be pressures to ever increase the accuracy. The cost of increasing accuracy increases exponentially to a certain point beyond which it becomes practically infeasible. This must also be clearly communicated to the business stakeholders and users to set the right expectations.

In the productionization effort for the typical AI product, the scalability features are a necessary foundation and tend to reoccur from one product to the next in more or less the same manner. These are therefore a foundation that can be repeated and so its investment cost can be amortized across a portfolio of products. The costs for increasing accuracy must be watched carefully as a request for a higher accuracy will almost invariably occur and this may put the product into an eternal science loop. That must be carefully cut at a scientifically reasonable point.

It is the features of user interaction that represent the main risk in terms of feature creep and thus budget and timeline. It is a well-known fact that typical users of most commercial software packages only use 20–30 percent of their features. Many features of most packages are used by a small minority of users. However, this majority of features generates the majority of maintenance costs. Frequent user requests for more features only increase this long tail in the distribution of use. Eventually many software packages become unwieldy—a common old-age disease of software known colloquially as **featuritis**. The complexity of many features ends up turning users away and is thus a common cause for the eventual replacement of this product for another newer, leaner product.

You should therefore be cautious about adding too many features and responding to all requests. Instead, you should only add the features that are needed by the majority and those that yield measurable value for the enterprise.

A Project's Journey

In the journey of a project from inception to product, you have seen the major stages with their decision gates—see Figure 7.1. Each stage has a few major deliverables that are collected in the chart. At a high level, this progress chart allows us to track the evolution of a project portfolio quite well.

Before the PoC stage, you plan the project as discussed in Chapter 6. This consists of identifying the stakeholders, conducting the discovery workshop, and filling out the canvas, which finishes in all stakeholders signing the final canvas document.

The PoC requires gaining access to enough data, conducting the science to prove the concept, and generating agreement among the stakeholders that the concept has, in fact, been proven.

The MVP needs the data and user interfaces to be built, agreement among the stakeholders that this does satisfy the most basic usability requirements, and then the deployment of this product so that it can be used.

The PoV stage needs us to identify the users that will conduct the proof process. They must then conduct the testing and document the value generated. Finally, you scale up the value to a full production deployment.

In production, then, you build out all the pipelines of data and secure them. The science is put on a firm footing. The interfaces to the users are fully built out. The product is deployed and users are trained on the software, including change management and feedback sessions to generate widespread user adoption.

FIGURE 7.1 The evolution of a project from design thinking to full deployment and maintenance

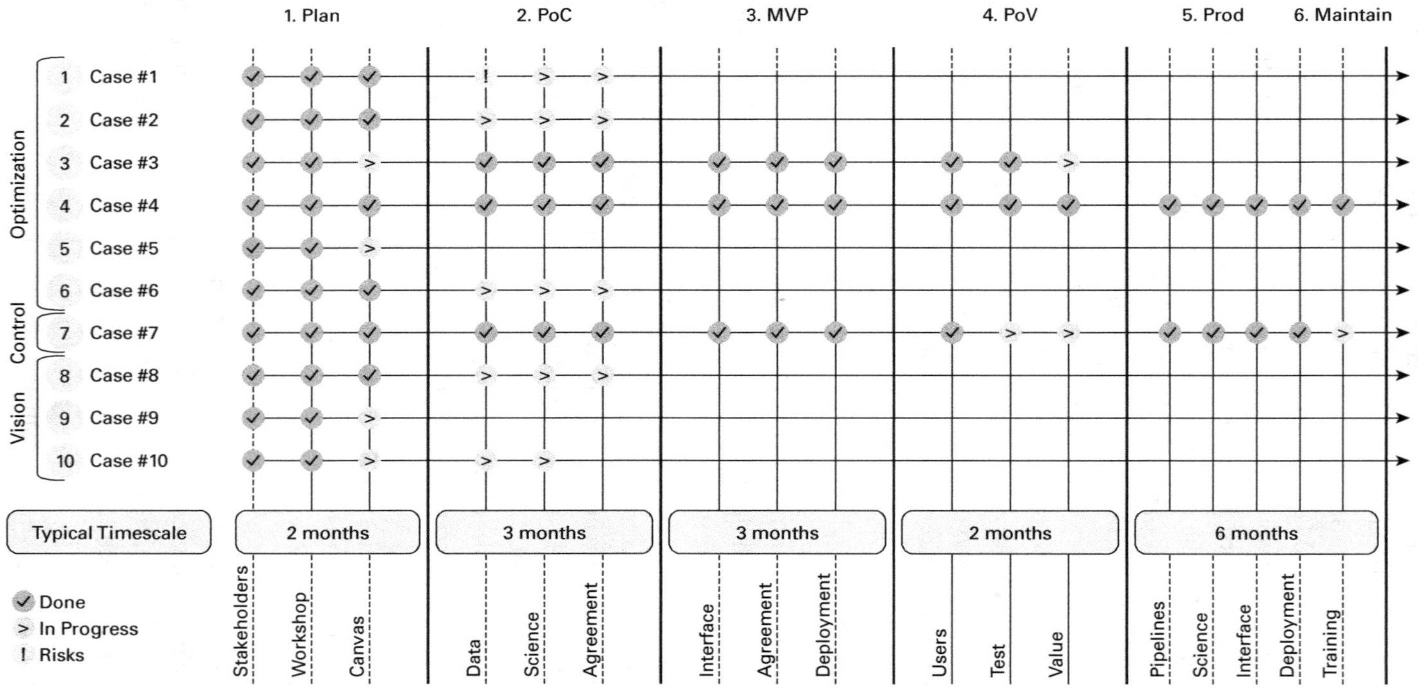

Finally, the project turned product goes into maintenance mode in which the product is continuously tracked. For an AI product, this takes on two main aspects. On the one hand, the product, like any other, receives user feedback and requests that need to be managed. On the other hand, AI models degrade in their accuracy due to the data drifting away from the training dataset's distribution. AI models must be retrained from time to time to maintain their accuracy and this is an essential part of maintenance.

The Idea of Agile

"Managers are people who do things right and leaders are people who do the right thing."

WARREN BENNIS AND BURT NANUS,
LEADERS: STRATEGIES FOR TAKING CHARGE

The idea of agile is closely related to the concept of an MVP, or minimum viable product. You could also think of an MVP as minimum viable planning. While this may sound humorous at first, let me explain why. Traditionally, software projects were managed like engineering projects for physical objects, using the waterfall methodology. This approach involves specifying exactly what you want at the outset and then building it according to these specifications.

For example, when building a car, you create components like tires, axles, and the body, and then assemble them to form the final product—see Figure 7.2. Only the finished car solves the problem of transportation from point A to point B. The customer is therefore unhappy until the very end.

The waterfall method relies on precise specifications and flawless execution to achieve the desired outcome. However, in software development, users, stakeholders, and engineers often struggle to specify their needs accurately. As a result, executing against initial specifications can lead to a product that is completely useless. This has been the experience of many software projects.

Instead, software development has adopted a step-by-step approach. Rather than focusing on building a car, you focus on solving the transportation problem. Initially, a skateboard may not be ideal, but it is more efficient than walking. A scooter, bicycle, motorcycle, and eventually a car represent incremental improvements towards the perfect solution—see Figure 7.3.

FIGURE 7.2 The waterfall method of building a car

FIGURE 7.3 The agile method of building a car

Physically, this makes no sense because a factory for making bicycles is not halfway to a factory for making cars. It is totally different and thus represents no progress. However, in the world of software, you have a great deal more modularity and flexibility than in the physical world. Solving parts of the problem may indeed represent true progress towards a full solution.

This approach is known as minimum viable planning. You plan for two weeks, then review your progress. You confer with the product owners and user group to determine whether your efforts are moving in the right direction. This iterative process, known as **sprints**, involves short cycles of development, feedback, and adjustment.

Agile project management relies on transparency, inspection, and adaptation. Each week or two weeks, the development team presents their progress to end users, who inspect and interact with the product. Users provide feedback, such as requesting additional features or restrictions. The team adapts to this feedback, planning for the next sprint based on user input.

These principles are operationalized in a framework called **scrum** (see Figure 7.4). Agile scrum combines the philosophy of incremental development and constant feedback with a structured framework. Scrum involves planning meetings, implementation, and review cycles. Each sprint lasts one to two weeks, and this iterative process has been observed to deliver useful results faster and with lower budgets than the waterfall approach.

FIGURE 7.4 The agile scrum framework

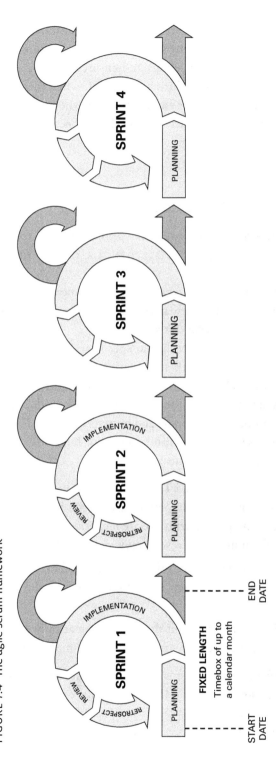

There are three major roles in scrum: the scrum master, the developers, and the product owners. The scrum master, essentially the project leader, runs planning and retrospective meetings and ensures effective communication. Developers execute the tasks during the sprint, while the product owners, representing domain experts or end users, ensure the project is on track and meets user needs. The product owner also defines when the project is complete and ready for testing.

User Stories and Artefacts

"Unclarified expectations are one of the primary reasons for broken trust."
STEPHEN M.R. COVEY, *THE SPEED OF TRUST*

In the agile process, the team creates a few documents known as **artefacts**. The minutes of the sprint review meetings are examples of these artefacts. A few more will be discussed in this section.

Rather than focussing on technical specifications, agile focusses on the human user of the product. All specifications therefore take the form of a so-called user story. The user story is a short, pithy, practical description of an activity that a user will want to perform with the product. It describes the scenario, the action, and the inputs and outputs involved. This description is usually just one sentence and can seem quite high level. Here are some examples to clarify what this means:

- As a control room operator, I want the AI to alert me when the drone flying on its automated flight pattern discovers a bird's nest on an electrical line and to supply me with precise location coordinates so that I can coordinate a maintenance crew to remove the nest.

- As a supply chain specialist, I want the AI to tell me how many of each type of item on a list to order so that I always have the optimal number in stock.

- As a machine operator, I want the AI to automatically modify the intake pressure once per minute so that the fuel consumption is at a minimum.

A product may have many user stories or may have only a handful. You do expect the user stories to be MECE: mutually exclusive but collectively exhaustive. That is to say, you want every possible activity or interaction to be captured, but only in one user story.

The stories are, of course, assembled largely by the domain experts and product owners with the guidance of the scrum team. Once they are done, the scrum team then assigns a number of points to each story in accordance with the estimated amount of effort that the team believes each story will take to implement. That estimation will include thoughts about the underlying technical requirements to make each story a reality. The points are a collective measure for the time effort, complexity, risk and uncertainty involved and so are a multidimensional estimate. Roughly speaking, one story point is equivalent to four to six hours of work effort by one person.

Armed with stories and points, the developers now start their work. In doing so, they produce work products. There are several work products that the agile methodology emphasizes because they form part of the process that helps the method achieve a good outcome in a short amount of time.

First and foremost is the so-called increment. That is the actual software product that is the point and purpose of the project. At the end of each sprint, the current status of the product—the increment—is shown to the product owners for feedback. This event is called a retrospective and will be discussed in the next section.

Another important artefact is the backlog. This is the collection of all those user stories that are not yet implemented. You always have a sprint backlog, which is the work to be done in the current sprint. You also have an overall backlog with the work to be done in the full project. Both backlogs are maintained by the scrum master and give an immediate overview of the status of the project at any time. Viewed from this angle, the final goal of the scrum team is to empty the backlog.

The backlog is a living entity. Stories are not just removed from the backlog when they are fully implemented. Based on product owner feedback, stories can be modified, and new stories may be added. This is the ultimate practical manifestation of the agile idea.

The agile method of software development started at the outset with two major practical observations. One was that users generally had a very hard time defining what they wanted or needed in sufficient detail to allow a software development team to realize it. The other was that users generally discovered what they wanted or needed during the creation process or as they were shown options. This turned a project into a moving target. With the waterfall method, this led to much longer projects than planned and many expensive change orders.

The software community then had the idea of simply forgetting about all the waterfall planning and showing intermediate work products to users all the time for feedback—and agile was born. Over many projects, the process of agile was refined and the community now has a well-established process for how you can achieve good solutions in a reasonable time.

The fulcrum of agile lies in the living backlog that captures user feedback in a structured manner.

Retrospectives and Planning

"Keep commitments ... is the quickest way to build trust in any relationship."

STEPHEN M.R. COVEY, *THE SPEED OF TRUST*

In addition to artefacts, agile has so-called ceremonies. The most important ceremony is the retrospective. This is the meeting that occurs at the end of each sprint. The development team shows the current increment to the product owners and obtains feedback. Based on this feedback, user stories are marked as completed, amended, or new ones created, and the backlog is updated accordingly. The team discusses what went well and what did not during the past sprint. The team then plans the next sprint. A few user stories are selected for implementation, and these are discussed with the product owners. Clarity is achieved on what is expected and the technical details are discussed and assigned to team members.

Another important ceremony is the stand-up. This occurs daily and may only take minutes. The purpose is to quickly coordinate what was accomplished the day before, what is to be done that day, and who will do what. This keeps everyone up to date, prevents duplication, and prevents gaps.

Sometimes, the artefacts and ceremonies can get too formal and they may appear to be a burden on the team and take too much time. If this occurs, you need to remember the spirit of the method. The spirit of agile is frequent adjustments to the feedback of the product owners to get to a problem solution as quickly as possible—knowing that you can only recognize a problem solution once it has been made manifest in actual software.

KEY TAKEAWAYS

1   Agree on clear and explicit expectations of the major stages in the project life cycle and the criteria that must be satisfied to move from one stage to the next.

2   Assemble the project team including the product owners, project manager, and the development team consisting of domain experts and AI experts.

3   Create user stories that collectively and exhaustively define in detail what the product will eventually do and start to work on them in the agile way.

8

User Experience and Interfaces

Value Generation Requires Stable Usage

"What if customers' single greatest need—ironically—is to figure out exactly what they need? If this were true, rather than asking customers what they need, the better sales technique might in fact be to tell customers what they need."

MATTHEW DIXON AND BRENT ADAMSON, *THE CHALLENGER SALE*

I often find that obvious things create the largest obstacles. By being obvious, you may not even be conscious of them or may not consider them worthy of discussion out in the open. What goes without saying may require a lot of saying, and loudly. One of those things is the simple fact that AI applications provide value to the enterprise only when they are used by users—regularly, stably, as part of their normal working day.

There are many intermediate states between no one using the application and stable usage. The application may be used either sporadically or by just a few users. Alternatively, the application may be used but its output is not heeded, or the right decisions or actions not taken as a result of its use. All three of these scenarios will yield failure—to deliver value.

A principal challenge of any AI project is to get all the intended users to use the application all the time and to act on its results.

Achieving this requires a combination of approaches that are part technical and part psychological. First, users must see value in the application *to them individually* in addition to the enterprise. Second, users should feel involved in the making and refining of the application. Third, the application must enable users to act on its results by fully integrating it into the various relevant processes of the enterprise.

Value to the enterprise is generally provided by lowering cost or increasing speed, productivity, capacity, or revenue. Value to an individual employee in the enterprise can be very different. They may be freed from an annoying repetitive task, the risk of making a mistake or getting injured could be reduced, their flexibility and freedom could be enhanced, they might be able to focus more on activities that are both valuable to the enterprise and enjoyable to them. Lastly, they may be able to improve their careers in some meaningful ways through this process.

Figuring out what matters to users is not difficult, but one must ask and take account of the fact that it will not be the same for everyone. Those reactions can be partly built into the application itself and partly into how that application is rolled out across the enterprise.

It is best practice to involve users as early as possible in the process of creating an application, ideally before you even decide to pursue the project. They then have the chance to be involved and to tell you what they need to be happy with it. If you ask them what they want and need, they will tell you. Sadly, many AI experts never ask the users about the experience they want to have when using the application. Users, like all humans, are emotional creatures. They are not motivated by objective metrics. All AI projects must therefore fulfill two sets of criteria: The objective business value to the enterprise and the subjective user experience to the users. This is not a chicken-or-egg question. The user experience must come first. The business value will then follow because the users are using the application.

Beyond this, users are typically anxious about either being replaced by AI or being blamed for AI's mistakes. Reassuring them that neither will happen is a crucial part of change management that we will address in Chapter 9.

Users See Only the Interface

"When people form habits or keep to rules, they are acknowledging that the costs of trying to optimize are too high. So they, in effect, decide not to decide. ... AI prediction is only useful if you are making a decision."
"When you are following a rule, you may be unaware of the value of gathering information and making a decision."

AJAY AGRAWAL, JOSHUA GANS,
AND AVI GOLDFARB, *POWER AND PREDICTION*

Every AI model or application has an interface for a human user of some kind. They also have interfaces to data and other software, but the topic here is specifically interfaces for humans, the user interface.

Interfaces could be quite simple. For instance, an AI model calculates the price for a flight ticket based on various information that you input into the search box. The interface to you is simply the output of a number on a website. A computer vision method tracking whether people are entering a danger zone can draw a rectangle around the person on the live-streaming image of that danger zone and provide some form of alarm. A chat interface can be used to communicate with other humans or language models.

Language models have been available in some form for a decade. They became famous with the advent of ChatGPT not because that model was a great deal better than others but simply because it was—for the first time in this industry—provided with a user interface that worked well for most people. It came with the same interface that everyone uses to send short text messages to their friends, namely a chat. Other language models previously only came with an application programming interface, which requires users to know programming languages. The revolution here was, in fact, the user interface and not the AI model.

Most users most of the time do not care about the fact that it was an AI model that produced the output. They just care about the output and the ways they can interact with it. If it is provided in a way that is useful and enables the daily work of the user without introducing any new frictions, it has a good chance of being adopted.

Being innovative, cool, or at the bleeding edge of technology matters to few users and, even when it does, wears off quickly and so cannot be a serious element of a rollout strategy. For generating value, AI applications must have the potential to become normal, boring, routine, and mundane!

Applications Must Be Integrated into Technical and Human Processes

"The primary benefit of AI is that it decouples prediction from the rest of the decision-making process, which facilitates innovation in organizational design via reimagining how decisions interrelate with one another."

AJAY AGRAWAL, JOSHUA GANS,
AND AVI GOLDFARB, *POWER AND PREDICTION*

The AI model that offers you the price of a plane ticket has a simple output interface, namely the display of a number. Once the model outputs its number to you, the model has been used. That display however does not yet offer value to the enterprise, in this case the airline. It offers value to the airline only if you buy the ticket. Actually, it only offers value if you buy the ticket such that, statistically, the airline makes more money than it did before having that AI model in place. It might have done that by correctly predicting desperation and getting you to accept a higher price, or by correctly predicting nonchalance and choosing to sell an empty seat at a cheap price.

In any case, the AI model's output is not the end of the value generation workflow. It is only a part of a more general process (in this case, it was the process of planning travel). In making a proposal for a flight, the website will provide you with a specific takeoff and landing time and information about the plane alongside the price. These pieces of information come from different sources and only the price was provided by AI.

Many AI applications are just like this. The AI output is meant to support either a decision or an action by a human. To enable this, the application must provide not only the AI model output but a variety of other information useful to the decision or action. Lastly, all this must be provided in a form that allows the human to easily, quickly, and securely make that decision or take that action.

The application therefore needs data input interfaces that go well beyond merely the input data required by the model. It also requires output interfaces that can interact with humans and, if needed, enable the desired action. In our example of plane tickets, the desired action is the purchase of the ticket, which needs the binding of a shopping system.

When buying plane tickets, the AI model that provides the price seems a minor element in a complex purchasing process. One may argue that we are not talking about an AI application at all but an ordinary software application that includes an AI model as one of its many components. This is the nature of most AI applications—the actual AI itself is a relatively minor part of the software system.

Realizing that most of the development effort into an AI application goes into the software wrapping, that allows both data and users to interact with the model and achieve value through usage, is crucial. Developing models and debating about accuracy is attractive, but it is the interface that leverages the model in achieving value.

In managing the corporate AI team, this fact is paramount and bears repeating often: It is the team's job to create business value by enabling a decision or action with AI—the work for which may be dominated by the enabling rather than by the AI. The team needs to think deeply about how to deploy and wrap AI models in reusable software layers that do the enabling because those layers determine not just the business value but also the cost and speed of creating these applications. A few examples will illustrate this point further.

When a computer vision model detects that someone has entered a dangerous area, an alarm must be released to notify someone of this so that they can do something about it. That might be a notification in a control room, or the sounding of a horn. Only after someone takes a timely and decisive action can people be saved and thus value provided.

When a price forecast for something you need to buy is made, this must be provided to the buyer such that they understand its significance. The price alone is not enough. It must be augmented by the likely amount you will need, your current inventory levels, your available storage capacities, and so on. You must also have the available cash to pay for it. Some of these may be simple lookups in databases while others may need AI models of their own, such as demand forecasts. All those elements need to be communicated in the right way to a person with authority to execute and then they must perform or defer the purchase action to derive value.

AI Predicts, You Must Judge, Decide, and Act

"If you are in the business of developing AI whose value is enabling decisions that are not being made, you will face an uphill battle in gaining adoption."
"Rules arise because it is costly to embrace uncertainty, but they create their own set of problems."
"Decoupling prediction and judgment creates opportunity."

AJAY AGRAWAL, JOSHUA GANS,
AND AVI GOLDFARB, *POWER AND PREDICTION*

The output of an AI model is called a prediction. This is an unfortunate word really, because in many people's minds this word suggests the meaning of a forecast, the meaning of something that happens in the future. For most

AI models, the output—the prediction—has nothing to do with the temporal future at all. Nonetheless, this is the word that is used.

The prediction, as the result of an AI calculation, has an associated uncertainty. That fact is probably the single most important characteristic of AI. The uncertainty is inherent. A mistake by AI is not a bug to be fixed but an event that should be made rare. The likelihood of a mistake can be reduced, but never to zero!

If the prediction is a number, the uncertainty is a range of numbers. The output might be 5.1 and the range might be between 4.9 and 5.3. Whether that is a small or large range, whether that range is good or bad, depends on the problem. For example, if we are forecasting the price to be $5.1 and we know this to be accurate to plus-or-minus $0.2, we might not care about the uncertainty and proceed under the assumption that $5.1 is right, approximately. However, if it is accurate to plus-or-minus $1.5, we might consider that prediction as effectively nonexistent because it is too uncertain to have any meaning. Therefore, we cannot use it as a basis for any decisions or actions.

If the prediction is a categorization such as "this belongs to group X" or "this is dangerous," then we have a probability of this being correct, which is necessarily less than 100 percent. That probability will inform us as to whether that prediction can be relied upon. What probability is good also depends on the case. For example, the prediction "you have colon cancer" based on the AI analysis of a colonoscopy video should come with an extremely large (~98 percent or higher) probability of being right to be taken seriously because it comes with significant consequences. However, the prediction "you will like this movie" can be output with a much lower chance of being correct because the consequences of being wrong are minor.

Most AI applications output their predictions without an indication of accuracy. I consider this irresponsible. Users are generally not aware which elements on the screen are AI calculated and which are looked up in a database of facts—and they should not have to be. The application ought to tell them in no uncertain terms that this number is AI calculated and has a certain range of validity. As AI experts, it is our responsibility to inform our users optimally and honestly about what is going on so that they can make up their own minds about what to do with it.

Putting these elements together then, AI provides a prediction with an uncertainty indication, the application provides data about the context of the situation. All this is in service of you. You are now tasked to do three things—to judge whether all this is informative enough to be a basis for decision making, to decide what to do, and to do it: or not.

To judge and decide, the application must provide the right and complete information with the required precision and provide it in a form that enables you, the user, to absorb it and conclude that this is a suitable basis for decision-making. The decision-making may happen in your human brain, but the decision itself must then generally be entered into some process, which may or may not be a software program. We expect at this point that the application will make this as easy and seamless as possible, being integrated into the upstream and downstream software landscape.

It is ultimately your action that provides the value here and so it is the user interface provided to you and your user experience that is responsible for the long-term value generation through the AI application. Providing great user experience is paramount and should be the top criterion for every AI application developer.

Again, by and large, users do not care how the answer (prediction) was arrived at—they care that this answer is correct, precise, and actionable through context and they care that it helps them personally. That it also helps their company is important, of course, but users do, naturally, care about themselves and their own experience first.

Measuring and Improving User Experience

"The view you adopt for yourself profoundly affects the way you lead your life. ... Believing that your qualities are carved in stone—the fixed mindset—creates an urgency to prove yourself over and over. [...] This growth mindset is based on the belief that your basic qualities are things you can cultivate through your efforts, your strategies, and help from others."

CAROL DWECK, *MINDSET*

Seeing that user experience is so important to AI value generation, how do we determine the quality of the experience? The clear answer is to ask and listen. Then to make the improvements suggested. What is implied in those obvious statements is profoundly important however: You must have the people and the time to do this and so budget for it up front in project planning.

Improvements can be made in phases, of course, and are often done over a longer period. For minor improvements, that is fine. Early in the lifetime of a new product, however, there will be some suggestions that are critical for acceptance. There are also some suggestions made by critical people—so-called influencers. These are not your famous social media influencers but people at the company whose opinions are respected. Whether they are supportive or not matters a great deal to get acceptance. You must determine, early on, who they are and keep in contact with them to make sure that they get what they need to be supportive.

Having determined who the influencers in your user base are, you will want to ask them about their experience. While the product is being built out in the MVP phase and later in the early productionization phase, these people should have an active voice in the design of the user interface and workflow. This not only gets your first version to a stage where it has good chances of offering a good experience, it also gets the important users bought into the new product.

Here are some of the questions you might ask.

How quickly do you need the information updated? Considering two timings is often important. **Cadence** is the amount of time between two consecutive updates of the data displayed in the application. **Latency** is the amount of time that elapses between the input data being generated at their source and the output of the application being available to the human user. Often, both timings have an effective cut-off so that if the application is slower than that, the application becomes useless. If the application is faster than the cut-off, this may or may not offer additional value. This is a popular trap for AI experts and software developers. They will enter into a competition with themselves to make the application as fast as possible, not taking into account that as soon as it is fast enough, any further speedups are useless and just waste time and resources. Figuring out what the required timings are and what influence they have on the usage of the application in real circumstances is important for the value generation but also for gauging the interest level and annoyance of the human user. For instance, no one likes to look at a spinning icon for very long.

How should the application communicate the accuracy of the AI model so that you have trust in it? No model is 100 percent accurate. For a specific output, we need to communicate the model's confidence in its own output. We may provide a probability that this categorization is correct. We may provide the confidence interval for a forecasted number. We may provide a graphical bounding box for image detection. In aggregate, statistical assessments of accuracy may be output. In such circumstances, it is important to review with users whether talking about concepts such as F1 scores, precision, recall, mean-absolute-percentage-error, and so on is understandable to the user population and offer a usable indication of whether this output is respectable. Ultimately, this is about communication to generate trust on behalf of non-AI-expert human users—the outcome is a distinctly emotional one.

What information does the application need to display to allow you to do what you need to do? In addition to the output of AI models, the application may need to provide a variety of other information to enable the user to accomplish a task. Often the task for the user is to make a decision. That decision-making process may need contextual information that is merely looked up and output. The ability to look those things up may not be trivial from a software engineering perspective, however.

How should this information be communicated? All the information, whether AI generated or looked up, must be displayed to the user in a way to enable the decision-making or action process. That may be graphical or just a list. Discovering the optimal way to display it is often a long conversation that may go through several trial-and-error loops. Knowing this, it is often useful to construct mock-up designs such as digital wireframes or pencil-and-paper drawings. The aim is neither to be fancy nor to win a design award. The aim is to be maximally useful. For that it helps to talk through the process that the user follows in great detail using several concrete examples so as to "experience" the way this would work and to iterate until the AI application supports that workflow in the best possible way.

How do you want to interact with the application to enable further process steps? Something follows after the AI application is finished. For the application to be useful, that needs to be known and analyzed. The interface that represents the transition from the application to whatever follows is the most crucial part of all. This interface can either make the application valuable to the company and enjoyable to the user, or the opposite. In fact, that interface is more often than not the human user themselves.

What features should be added to make you more effective and efficient? Open-ended questions are most likely to provide you with good ideas for making things better. While many good ideas will be generated, most of them will provide little aggregate value as they may be desirable to just a few users and may be difficult to realize. These ideas must be treated with some suspicion as the primary cause of death for all software applications is featuritis—the disease when a software has so many features that the cost of maintenance of the application exceeds license revenues. It is anecdotal in software that a majority of features are used by a minority of users. For an enterprise AI program developing applications for in-house usage, it is crucial to keep maintenance costs low, and so features should be kept to a minimum. The goal is making the application good enough for practical use and making the users happy enough to efficiently use the application—not to fulfill their every desire.

How can the application be better? Another catch-all question yielding good ideas as this moves the attention away from the user to the application itself and the outcome. The user can be reminded that the application must fulfill two goals. For the user, it must offer good experience, and for the company, it must offer business value. This addresses the overall goal of value.

If you prefer to ask a larger number of users to collect statistics and so cannot have an open-ended conversation with individuals, you can consider doing a survey. There are multiple standard differential surveys available for use so that you do not have to design your own. These standards are helpful not only in saving you some work but in enabling comparisons to an established baseline reference point from other companies and products. The simplest form of feedback is to measure the net promoter score: On a scale from 1 to 10, how likely are you to recommend this product to a colleague?

KEY TAKEAWAYS

1 Applications must be used to generate value by human users who only see the user interface.

2 Users care about their own subjective experience while using the application and about how it is integrated into the workflow process that they engage with.

3 Measure and improve user experience by asking questions and actively listening to the opinions of influencers who can sway the opinions of the whole user base.

9

Change Management and Adoption

Adoption

"The way you are conditioned to see the world in your own culture seems so completely obvious and commonplace that it is difficult to imagine that another culture might do things differently."

ERIN MEYER, *THE CULTURE MAP*

There Are No Unhappy Users

So, who is going to play with the amazing tool you have created? How will the AI application you've developed be adopted and actually used by real users? This is a key issue. The number one reason AI projects fail to deliver business value is that the intended end users simply do not use them. This has nothing to do with mathematical methods, algorithms, accuracy, technology, software, or cloud infrastructure. The primary challenge is user adoption. Therefore, as leaders, managers, and organizers of AI projects and tools, you must focus on the change management required to ensure adoption.

To use AI effectively, you must convince people that it is in their best interest to use it, that it won't harm or threaten them. Ideally, you should involve people very early in the process of creating the tool to ensure they are engaged and satisfied, and that you are developing a tool that works for them.

There are certain prerequisites for adoption. While these do not guarantee success, they are necessary conditions. First, having business champions and executive leadership support is crucial. Senior leaders must back both the problem being solved and the tool used to solve it. Involving user groups early in the process is also essential. While you may not be able to involve every end user if the group is large, you should include representatives who can provide insights into real-life scenarios to ensure the tool fits their needs.

The second prerequisite is a crystal-clear definition of the problem. This is not a scientific issue but a real-world challenge experienced by users. You must define it as precisely as possible. Getting all those stakeholders to agree to that definition is your first project milestone and a genuine achievement.

The third prerequisite is integrating the AI into the larger workflow. You need to understand where the AI application fits within the broader process, which may involve technology, software, physical equipment, and human decision-making. Defining the boundaries between the AI application and the larger workflow is critical to ensure the tool is adoptable and delivers real value. For example, if you provide a chatbot to field technicians who spend their day driving trucks and dealing with machinery, typing into a mobile phone is unrealistic. Enabling a verbal interface that allows them to talk to the chatbot and receive responses through a speaker is essential. This is not advanced AI but a necessary user interface feature to ensure adoption.

Education is another critical component. Both the AI team and the domain experts must educate each other about their respective fields to avoid misunderstandings. This two-way education helps both parties understand each other's language and worldview.

Finally, you should avoid doing AI for AI's sake. AI can offer great value in certain circumstances but may not be necessary in others. If a non-AI tool or an off-the-shelf solution can solve the problem, you should use that. Only when AI is necessary and developing a new AI solution is required should you proceed.

These prerequisites are covered by the design thinking canvassing process discussed in Chapter 6. This process ensures you are doing something that adds value to the company.

Remember that you're going through a four-stage process: proof-of-concept to ensure feasibility, minimum viable product to ensure usability, proof-of-value where end users demonstrate real experiential value, and, finally, production. If you reach the proof-of-concept stage and the concept is not proven, it is difficult but necessary to consider terminating the project. You need to determine whether the solution truly stacks up, solves the problem as originally posed, and is accurate enough for practical purposes. Both the scientific core and the graphical user interface must be checked to ensure they meet these criteria.

You must iterate with the champions and users to zero in on the right solution. There is a danger in a regular agile process of solving a problem that, step by step, you deviate from the initial problem posed. It's essential to stay focused on the original problem and ensure the solution addresses it.

By involving business champions and end-user representatives throughout the process, you keep everyone in the loop, allowing for valuable feedback and input, particularly for the user interface. This increases the likelihood that the application will be used.

Change as Threat or Opportunity

Now, let's discuss the psychology of change management. John Fisher's change curve (see Figure 9.1) illustrates the psychological reactions people have when presented with something new. People generally prefer steady states and resist change. The initial reaction is often anxiety, where individuals question their ability to cope with the change. This may be followed by a brief period of happiness, thinking that at least something is changing. However, fear soon sets in, worrying that the change may not be what they had hoped for.

People may react in one of two ways: denial, where they refuse to acknowledge the change, or fear of the impact on themselves, particularly the fear that AI will take their jobs. It's important to reassure employees that the goal is not to lay people off but to enhance their roles through increased automation, safety, and productivity. AI can eliminate tasks but not entire jobs, allowing employees to focus on more intellectual and less repetitive tasks.

After overcoming the fear of job loss, individuals may feel an organizational threat, realizing that they need to learn new skills and adapt to new tools. This can lead to a period of disillusionment as they try to figure out how they fit into the new reality. At this critical juncture, there are three possible reactions: disillusionment, hostility, or gradual acceptance.

Disillusionment leads to individuals ignoring the change and not using the new AI application. Hostility involves actively trying to undermine the application by pointing out its flaws. While this feedback can be useful for improving the application, it is challenging to manage. The middle path involves individuals questioning how they fit into the new world and gradually accepting the change as they learn about the new AI application and its benefits.

As a leader and manager, it is your job to paint a clear vision of the future (see Chapter 1 on AI strategy), showing end users how the new way of working will benefit them. By leading people through this change curve, you can increase the chances of adoption. Change management and adoption

FIGURE 9.1 The John Fisher change curve

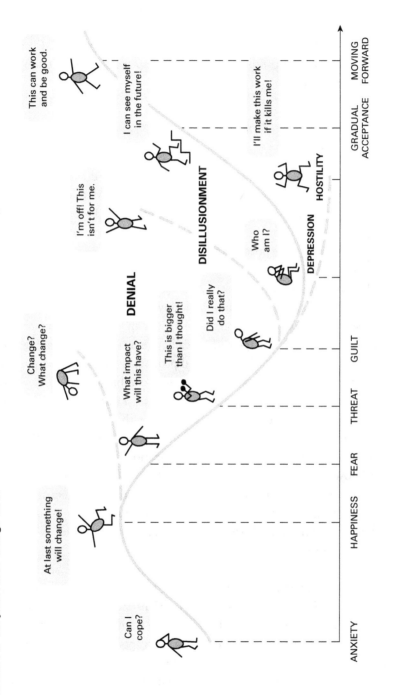

are critical for the success of AI applications. Reports by Accenture, Deloitte, Gartner, and other market research firms have shown that out of ten AI projects, approximately nine failed to deliver business value, and eight of those failures were due to inadequate change management.[1] Ensuring that your staff are comfortable with the problem solution and its implementation is paramount.

Primary Causes of AI Project Death

Creating business value is the ultimate objective for AI projects. However, there are three main reasons why they might fail to do so. Curiously, many non-specialists tend to assume that failure is due to data quality, data availability, model accuracy, and such technical reasons. In reality, these reasons are quite uncommon as there are established ways of mitigating those elements or ways of spotting them prior to starting the project and thus not causing it to fail.

The main reason is that users do not use the application. There are many subsidiary reasons for this, but it boils down to a simple refusal by a group of people to use your application. As such, it is an emotional reaction and thus an issue of change management. Ideally, this is solved by making the users part of the process by which the final product is partly created by them and so there is no reason for them to refuse it. A combination of early solutioning (Chapter 6) and project management (Chapter 7) will bring this about.

A second reason is that users cannot use the application. Even though they want to use it, the application is so poorly integrated into the human user experience, the other software applications, or the overall process and workflow that the users cannot use it. Most of the time, this is not a strict inability to use it but rather a problem of annoyance, speed, or comfort. Regardless, if users feel that it is sufficiently cumbersome to use, they will regard it as unusable. Carefully watching user experience will solve this aspect (Chapter 8).

A third reason is that you have solved a problem no one had. This sounds comical but occurs frequently, and I have been guilty of this myself on occasion. Most of the time this is caused by a misunderstanding between the AI and domain experts. When the groups met initially, they spoke the same words but understood very different concepts behind those words. For instance, the word "database" means a very precise software product for AI experts but may simply refer to any place where documents are kept to a domain expert. On occasion, I have heard people even refer to a physical warehouse with paper documents in it as a "database."

Additionally, both groups tend to consider certain things obvious. A few facts or ways of doing things may be so obvious that they are not even conscious to the experts in that field and are therefore hard for them to teach to others. As this happens from both sides, it is actually difficult to reach a point where both understand the other side sufficiently well to work on the same problem. Here too, the initial conversations yielding the canvas described in Chapter 6 are important in mitigating that risk.

If all three of these main reasons for AI project failure are carefully mitigated, the chances of success are fairly high.

Interfaces

There are two primary interfaces for any AI application. One is the data interface, which automatically obtains, cleans, and structures data from various sources behind the scenes. The interface you want to focus on is the graphical user interface (GUI), where a human interacts with the application, views the results, and makes decisions.

Closely associated with the user interface is the user experience, which encompasses the psychological response to the interaction. This includes how users feel about the interface—whether it is easy or hard to use, whether it speeds them up or slows them down, and whether it looks good or bad. These experiences are crucial for user adoption.

A common issue with user interfaces is that they are often designed by software engineers without considering the end user's perspective. This can lead to interfaces that are difficult to use and result in user frustration or refusal to use the application, regardless of its scientific value.

There are four main types of user interfaces for AI applications:

1 *Alert*: The application sends a message, such as an email or SMS, to alert the user. The user must then decide what action to take. The frequency and delivery method of these alerts can significantly impact the user experience.

2 *Dashboard*: A web page is populated with graphics, charts, and numbers that the user can check at their convenience. This is a passive interface where the user initiates the interaction.

3 *Instruction*: Also known as prescriptive analytics, this interface tells the user what to do. It integrates more deeply into the user's workflow by providing specific actions to take.

4 *Interaction*: This involves a two-way interaction between the user and the application, such as a chatbot. The user asks questions and receives answers, leading to a conversation.

You will see all four types of interfaces. While there may be other types, these four cover the majority of applications. It's important to consider the user's experience when designing these interfaces. For example, the difference between receiving an SMS message, a Teams message, or a notification bubble can affect usability, especially in environments with poor internet connectivity or bright outdoor lighting.

Integration into the workflow is also crucial. After receiving an alert or instruction, users need to take action. The messages must be practical and provide clear guidance on what to do, when to do it, and what tools or parts are needed. This ensures that the AI application fits seamlessly into the user's workflow.

Providing context is essential for decision-making. The AI application might need to include links to technical documentation or maintenance manuals to help users make informed decisions. This means that an AI application is not just an input–output interface but may require additional integration to provide the right message.

In summary, a happy user is a user. An unhappy user is not a user. Ensuring a positive user experience should be top of mind when developing AI applications to encourage adoption and deliver value.

Decision-Making

AI products often support a decision-making process. It is helpful to go over that process in detail before even considering building an AI application. In this section, this will be illustrated by a concrete example of maintenance of a physical machine such as a pump, compressor, or turbine. These machines sometimes stop operating spontaneously. Finding the right spare parts and qualified repair personnel can be time consuming and so the field of predictive maintenance attempts to provide a forecast of such failures a week or two beforehand so that the maintenance effort can be planned. To be useful, this alert must provide the time and reason for the failure and must be both accurate and reliable so that we can be assured that, more often than not, the activity will fix the problem before the failure occurs. That being said, Figure 9.2 is the logical tree of possible events and decisions.

FIGURE 9.2 Decision tree for turbine predictive maintenance

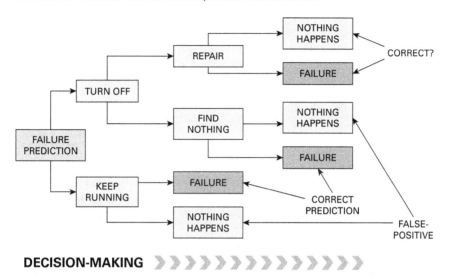

DECISION-MAKING ≫ ≫ ≫ ≫ ≫ ≫ ≫ ≫ ≫ ≫ ≫ ≫ ≫ ≫

The process starts with a predicted failure. What now? You have two basic options: to either turn the machine off or keep it running.

If you turn it off and inspect, you will either perform a repair or find nothing.

If you repair it, nothing happens, or a failure happens. If nothing happens, you have no evidence that your model was right or that you did the right repair. If a failure happens, you also don't know whether your model was right because you affected the system, but you definitely get what you wanted to avoid, namely the failure. Both are bad outcomes.

If you find nothing and turn it on again, nothing happens, or a failure happens. In the case where nothing happens, you know that AI produced a false positive and that is not good. In the case there is a failure, you know that the AI was right, but you ended up with a failure anyway. Both are bad outcomes.

If you keep it running, the machine will either fail or nothing will happen. If it fails, the AI was right, but you still have a failed machine. If nothing happens, the AI was wrong.

This brief analysis shows the difficulty of using AI in practice. Note that nothing was said about how accurate the model is. Knowing, for example, that the model will get 10 percent of its forecasts wrong will only add fear and doubt to all this. Ultimately, this analysis is the reason why predictive maintenance systems do not sell well in the process industry.

The only way to get out of this problem is to have a senior executive make the decision to always go with the turn-off and repair path and to then take the statistical risk that sometimes failures will still happen. It is then only after a statistically relevant number of instances that we can get together and decide whether or not AI really made the situation better in terms of reduced downtime of these machines across the fleet.

If the executives approach this problem from the point of view of a proof-of-concept where one or two cases are analyzed, the discussion will invariably end up with an analysis similar to the above and the use case will never be pursued.

It is absolutely crucial that you have these discussions before starting out on your AI project. If a business has not thought through what follows once the AI model is made and how they are going to use it, there is no sense in building it. The world is full of scientific papers where AI models were built and never used despite being accurate and applicable. In most cases, the wasted effort was preventable by a careful conversation and thinking through what would be done by the end user.

Leading the Change

"Leadership is scarce because few people are willing to go through the discomfort required to lead. ... If you're not uncomfortable in your work as a leader, it's almost certain you're not reaching your potential as a leader."

SETH GODIN, *TRIBES: WE NEED YOU TO LEAD US*

In Chapter 7, the major stages of an AI project were introduced at a conceptual level. This section will revisit them from a change management perspective, with the clear goal of adoption and value generation in mind, and (humorously) connect them to the well-known five stages of grief.

Denial at Planning: This Will Not Work

When planning for a new AI project, change management begins at the same time. It is clear that an AI project will cause change eventually as the very purpose of AI is to automate some piece of a process. What will change, in what ways it will change, and who will be affected are largely determined right at the start when defining the scope.

The majority of AI projects that I have personally seen or heard about in some depth across many industries have started with a statement to the effect of "This is just a proof-of-concept so that there is no need to involve [insert long list of names] or to reserve a large budget." Those are generally the words of doom for the project for three reasons.

First, it sets the expectation that this is just a proof-of-concept that will finish after feasibility is demonstrated. In clear terms this means that the project will never lead to a product and so no one will ever have to do anything different—apart from perhaps being loosely aware that some concept has been demonstrated. There will be no change. You are safe.

To be fair, in most cases, this statement is correct as most of these projects really do not go beyond the PoC stage, although, in my view, this is largely due to this statement being a self-fulfilling prophecy. Setting up the project as an academic exercise is motivated by not wanting to draw much attention to it early on to save the organizers some corporate politics before solid evidence is not in hand.

Second, it excludes the people mentioned, who are often significant stakeholders. Most often, these include senior management and end users. The project team does not want to include senior management for fear of failure and does not want to include end users for fear of backlash and arguments. However, if senior management is not involved, the project is unlikely to get support. If end users are not involved, the outcome is unlikely to be used—and this is the death knell of any new endeavor.

Third, PoCs are generally constructed to be cheap. That makes sense as everyone is trying to hedge the risks. However, if all the funding you have is for the PoC, what will you do if you succeed? In the best-case scenario, you will now have to lobby senior management for budget. As they do not know about the project and budget cycles are long, this will lead to a significant delay, during which crucial people may leave, or crucial knowledge will be forgotten. The loss of knowledge is particularly likely as PoCs are rarely documented well. In the worst-case scenario, this project vanishes despite academic success.

When vendors are involved, it gets trickier. Common practice has vendors heavily discount PoCs in order to easily and quickly gain access to the much larger budget for the "real" project. The vendor may not realize that the budgetary decision-makers are not even involved, let alone have reserved a larger budget for the follow-on project. Going further, vendors are often reluctant to even inform enterprise stakeholders about the true cost and duration of the full project beyond the PoC, fearing that it will not go ahead. Misunderstanding the complex interaction of enterprise budgeting and internal politics between stakeholders can hamstring a project before it even starts.

If you want to drive business value from AI, you must develop a full AI-enabled software product that is used by end users and supported by budgets created by senior management. It is imperative to set up a project with that expectation to those stakeholders.

When the stakeholders come together for an initial meeting, there are usually a few who believe this will not work. While that may indeed be the case, you should set the project up for success by gathering support and resources assuming it will be a success. The message is that the project will *begin* with a PoC, but it is not *limited* to a PoC. If the PoC turns out to be a failure, no material damage is done because the extra budget has not been spent. But the lessons learned get widely disseminated—improving the odds of the next project.

Anger at Proof-of-Concept: This Should Not Work!

Whatever you do, there will be stakeholders during the PoC who are hoping that it will fail. They do not want the work or the change. They see AI as some fancy math thing that does not understand their world. While it may understand someone else's world, *their* world is far too different, unique, and difficult for an AI to understand. After all, they have done things like so for a very long time and are getting by just fine, thank you very much. Plenty of outsiders have come before, tried to shake things up, and gone home with egg on their face.

The moment the PoC does succeed, you have two principal problems: The stakeholders will debate whether or not this is really a success, and whether the project should proceed.

All projects are troubled by unclear definitions of "done." It is important to define, at the start, what success will look like. Ideally, this is a numerical threshold on some variable that is objectively measurable. This is an insurance policy against people trying to argue away from success at this moment. The canvas process illustrated in Chapter 6 is partly aimed at this problem by trying to achieve clear agreement among all stakeholders on what is to be achieved, and documenting this agreement.

The people who were hoping all along that you would fail will now attempt to change the definition of success or argue that the project should not proceed despite success. Common arguments are that the definition of success was too lax, that it will cost more than projected, that its business value is less than previously thought, that the problem itself no longer exists, that there is another solution available now, or that the users will boycott the application. You must have responses ready for these situations. Again, having shored up group agreement in advance makes it harder to make any of these arguments.

Most of the time, these are just rationalizations for the real—emotional— reason for not wanting the project to go ahead: Fear of change. When replying to the rational arguments, it is good to pay lip service to the argument but to mainly address the underlying reason and reassure people that the change will be managed well. At this point, the most common aspects of the fear are that the AI will take away their jobs, that the AI will create stress, or that the people themselves cannot cope with the fancy new technology that AI represents.

You must assure them that AI will not take their jobs and that the software development life cycle that now starts will take into account their needs and user experience to remove stress and make certain that all of them can use the new product with ease. Their voices will be heard and taken seriously.

Bargaining at Minimum Viable Product: Let It Not Work

Having demonstrated that the idea is scientifically viable, the model is now wrapped in a minimum layer of software providing just enough usability for the AI model to be testable by an enthusiastic hand-picked alpha tester. The MVP consists of a working data input stream, the model from the PoC, and a bare-bones output to the human user.

The phase of turning a science project into a workable, albeit basic, software program is not easy. The MVP does have to have enough functionality to be usable and so several challenges must be navigated so that the data arrives at the model and the output arrives with the user. This usually surfaces a host of IT and software engineering challenges that require both work and political maneuvering as access privileges need to be sought.

It is not uncommon that a challenge only becomes visible after the prior challenge has been solved and so the project cannot be planned from the start—see Chapter 7 on agile project management. This is comparable to the infamous "fog of war" where you cannot see clearly and must make decisions under conditions of uncertainty.

Realizing this, some project participants secretly or publicly hope that the attempts will fail. This can mostly be seen by one of two attitudes. The first type of doubter will express doubts, clearly stating that this is likely not to work, that there are so many problems, and that further problems await down the line. The second type appears to be helpful, pointing out that one might try this approach, that technology, some software, or suggests calling an expert for more input. These are delaying tactics consciously or subconsciously designed to cost time and resources.

All projects, even agile ones, are bound by some resources such as time, budget, and executive patience. As discussed in Chapter 10, AI projects in particular are also challenged by opportunity cost as there are so many projects on the roadmap that a dragging project puts multiple other projects in jeopardy.

While the PoC was mainly a scientific challenge where the domain experts had to weigh in significantly, the MVP is a technical software engineering challenge in which the domain experts cannot really help. For this reason, it is best to keep the circle of participants in the MVP tighter than in the PoC and to watch for the above-mentioned attitudes. This tight circle might include the AI experts, any required software engineers, a user-interface designer, and the one or two key users who will use the MVP once it is ready.

You can further mitigate the risk by using the platform strategy discussed in Chapter 4 whereby you keep the development effort of the MVP to a minimum approaching zero because all the building blocks needed have already been built and are available in the form of a general deployment framework just waiting for a new model.

Once the MVP is built and you show it to the business, you are likely to hear complaints about how it does not do something, is clunky, looks bad, and has various other deficiencies. Your audience is judging from the viewpoint of the ideal final state. It is necessary at this point to remind the audience, often and loudly, that this product is *minimum*. Better versions will appear, but later in time.

Depression at Proof-of-Value: It May Work

During the MVP phase, the project team and the business team worked together to select one or two alpha testers. These are members of the pool of end users. Ideally, they are enthusiastic about the product, have weighed in on how the MVP was built, and have the status of influencers in the pool of end users. You want the other end users to take their opinions seriously. As they were part of the MVP phase, they know the product well and fully understand the use case for it. They are thus also able to judge in which scenarios they can use this prototype with the highest chance of success.

The proof-of-value is the activity in which the alpha testers use the MVP in a few real-life cases. The cases must not be constructed or simplified in any way. The test ought to be fully realistic. While the alpha testers are expected to document their experience fully to provide helpful suggestions to the developers for improving both the model and the software, the real emphasis in the PoV lies on determining the business value.

Initially, before work on the project even started, the stakeholders had a thesis on how much value the future application would provide. It is at this point that this is verified.

For some stakeholders, this is the stage where they psychologically turn to a belief that this may work out after all. It is still a depression mode because they are still fearful of all the changes that will need to happen if indeed the value materializes, but they can see the light at the end of the tunnel that the optimists may have been right all along. There is some hope left of killing the project by either insisting on testing the MVP under onerous conditions or setting an unreasonably high bar on the value judgments provided by the alpha testers.

In many projects, the data that makes up the proof is scrutinized carefully, and some people try to discredit the proof. You are served well if you prepare for this stage well in advance—in cooperation with the alpha testers—to choose cases that are difficult to argue with and to document the actions and decision taken with great care. The more forensic the evidence, the more confidence your jury will have to declare the PoV as a pass.

Passing the PoV stage is a big deal because this means that everyone has acknowledged that a real product delivers real benefits and so this effort has now turned into a product that legitimately exists.

Acceptance at Production: Sigh, It Works

Critics along the way now transition to acceptance mode. This thing is real. The product now enters the production phase. While this does not yet mean that the product is available for regular use, it means that the MVP will now be built out to get there. All these stages are needed to weed out projects that do not work because productionization is the costliest step in all resource categories.

Many stakeholders reluctantly accept that the product is now real. They realize that the change is coming for certain, but they still do not like it. Clarity is helpful by providing a plan on the timing of the product's release. Advanced training to the end user group in what the product is meant to achieve, the changes it causes, and the process elements that remain untouched will calm the uncertainty felt by many. Depending on the complexity of the product, this training could be a single session and take little time. Most AI applications are very simple for end users to deal with as virtually all their complexity is internal to the data preparation and AI models. When you release the product, the audience is prepared and anticipates it. Sharing some of the documentation from the PoV can be helpful in letting users know in what ways the software has been found to be valuable.

As in most things, starting by explaining why the company is doing this is the key to gaining adoption. If people receive the product in the spirit that corporate is forcing them to accept another tool they did not need, then you are starting from a difficult point. Communication that there is a good reason for this, and that there are benefits for the individual user, is great. Putting this communication out there in advance of release is even better. Do not explain yourself after the fact as an annoyed audience may not hear you out.

Feedback is crucial. Stakeholders and end users will want to provide feedback. They may or may not proactively do it. The responsibility of getting feedback is on you, not on them. While you cannot force anyone to give feedback, you can proactively create channels by which they can provide it and encourage them with an open mind to participate. Short-message channels, office hours, or training sessions with question-and-answer sessions are some options for creating a space for providing feedback.

Having collected it, you must respond. The best way to do so is to communicate clearly and publicly on what the items are, which ones will make it into the product's roadmap, and which ones did not make it and why. People can now see that the product will evolve—this provides hope that the things they do not like or perceive as missing will be fixed. Most importantly, it shows that you heard them and take the feedback seriously.

In conclusion, managing the change and generating adoption is not easy and neither is it quick. This requires time and personnel. It is paramount that these activities be part of your time and financial budget plan from the start and that whatever executive's approval you need to get is fully aware of this. Change management is often seen as a fluffy, touchy-feely optional side activity that we can do if we have time and money left over (which of course we never do). Actually, it is the fulcrum on which everything turns because if users do not use the product, the entire process has been in vain as no value will be generated.

From the point of an enterprise that is not directly in the AI business, software and AI change management is often regarded as easy compared to the changes closer to the core activities of the enterprise, which are often more physical in nature. In reality, changes to a machinery park, a building, or an industrial plant usually affect a limited number of people and can be worked around with little advance notice and soon accepted as a "new normal." Software changes often affect more people, have more moving parts, pose more opportunities for error, and so are more complex.

KEY TAKEAWAYS

1 Most AI projects fail to deliver value to the enterprise due to one of three reasons: Users refuse to use the solution, users cannot use the solution due to a poor user interface, and the solution solves a problem that the users never had.

2 Adoption is generated by providing users with software that is well integrated into the software landscape and the human processes in place alongside a well-curated training process in the new models and tools.

3 Change management is an essential activity that needs time and budget at every stage of the process to handle the legitimate and non-technical barriers that inevitably arise due to fear of change.

Note

1 Accenture (2025). Making reinvention real with Gen AI. www.accenture.com/content/dam/accenture/final/industry/cross-industry/document/Making-Reinvention-Real-With-GenAI-TL.pdf (archived at https://perma.cc/U9JM-ZXGX)

Deloitte (2025). Now decides next: Generating a new future. www.deloitte.com/content/dam/assets-zone3/us/en/docs/campaigns/2025/us-state-of-gen-ai-2024-q4.pdf (archived at https://perma.cc/FYJ3-SHUA)

Boston Consulting Group (2025). From potential to profit: Closing the AI impact gap. https://web-assets.bcg.com/0b/f6/c2880f9f4472955538567a5bcb6a/ai-radar-2025-slideshow-jan-2025-r.pdf (archived at https://perma.cc/J63S-R2UB)

Kearney (2025). Are CEOs ready to seize AI's potential? https://info.kearney.com/5/8849/uploads/2025-ceo-ai-study-final.pdf (archived at https://perma.cc/K83F-85BM)

Gartner (2025). A journey guide to deliver AI success through AI-ready data. www.gartner.com/document-reader/document/6709934 (archived at https://perma.cc/5EVR-FJWT)

10

Managing Costs and Rewards

How Much Will It Cost?

"The truth does not come without a tax of effort."

HERCULE POIROT IN *A HAUNTING IN VENICE*

Next to "How long will it take?" the most popular question from all sides is determining the cost of the project. While there are many components of the cost, business stakeholders want and deserve two simple numbers as an answer. There is a one-time cost to make the application and a recurring cost to maintain and run the application. In both these categories, a distinction between labor costs and non-labor costs is important. The next subsections discuss the costs to make and run AI applications, and raise some thoughts specific to labor costs.

Costs to Make AI Applications

Clearly an AI application needs an AI model. The model can either be purchased for a known fee on the market or must be created inside the enterprise. If you can purchase it, it probably makes sense to do so and then the cost is known. If you need to make your own models, this represents a cost to be discussed in the next subsection.

Beyond the model itself, you will need to create input and output software pipelines for the data. These pipelines must not just transport the data securely, robustly, and quickly but also possibly transform the data and clean it. While data cleaning before model training is often a manual activity on a large amount of data, the live data flowing into a model must also be clean—otherwise the application will be an example of garbage-in-garbage-out.

Live data cannot be cleaned manually, and so software rules are created to do this that are often problem and data specific. That is to say that these pipelines are not trivialities and represent a problem-specific work effort.

A user interface is also required. This could be simple like a notification or could be complex like a full software program with many features. It depends on the needs of the users and must be considered.

Apart from the end users, the application typically outputs various information to the admin users and developers to inform them about the use of the application. This can be simple, such as the writing of log files that are checked only if there is a need after a malfunction of the software.

IT will be involved in making AI applications to provision resources, handle cybersecurity, enable user authentication, and provide access to systems and tools. Applications will need to meet standards and be reviewed before deployment to make sure that they meet those standards.

Many of these items will be needed for any software development project. For AI specifically, we now need to add model observability. Once trained, the model is static but the world that generates the live data being streamed through the model changes. Eventually the distribution of the live data will differ significantly enough from the distribution of training data that the accuracy statistics of the model are no longer good enough (this is known as data drift). At this point, the model must be updated with fresh training data. The task of watching out for this moment and gathering the necessary novel data in the meantime is called observability.

In addition to data drift, observability will watch performance issues such as latency—the time elapsing between data acquisition and delivery of the AI output to the user—and trace the processing piece of the application for the various elements of consumption costs, with the aim of detecting possibilities for saving money and resources. For example, if a camera is installed on a forklift to prevent it from hitting pedestrians, the latency must be in small fractions of a second, allowing the brakes to be activated automatically in time to save the person.

Apart from these technical matters, project management is needed to keep the technical matters aligned with changing business needs and perceptions, as well as with the requirements of IT and others. Change management is needed, as discussed in Chapter 9, to make sure that the users are involved and ready to use the application once it is completed. Finally, users need training and support when they are ready to use the application. Those items are responsible for considerable human labor costs.

All these items cost money. It is impossible to say how much actual money any of these will amount to but must be estimated on a case-by-case basis. The point of enumerating them at a high level here is to say that there is a great deal more work than just the scientific work of making the model. There are many more people involved than just the data scientists who make the model.

Vendors submitting commercial proposals for AI models and applications generally include only few, if any, of these elements and assume that the enterprise will supply them. They are often under pressure to offer a small price tag and so reduce not only some of the associated expenses but even core features. Vocabulary that points in this direction includes proof-of-concept, minimum viable product, phase one, or similar. Business stakeholders must be made aware of the fact that the total cost of building the application is going to be between double and five times the cost quote of the vendor.

Internal data science teams are in a similar position. They will make an AI model and then expect it to "just" be productionized, not being aware that approximately 80 percent of the total building costs of an AI application go to the various activities previously listed. This fact raises eyebrows not only due to the financial cost but also due to the cost of time and personnel. If these activities were not budgeted for or were not considered in the business case made for these applications, the awakening might be rude.

Practically speaking, I recommend making a list of all the IT tools and resources you will need for the project and thinking through what the resource charges will be for them. It is likely that most of them will be "free" for the purposes of this project because the enterprise has already licensed them and there is no incremental cost. Application development, however, may generate significant cloud consumption costs.

For the work items, I recommend tabulating how many person-months or person-weeks are likely to be necessary for various of the above items, to note which team will supply them, and what skillsets are necessary. Here is a brief example:

1 Database setup. Database administrator from IT. One person. Two months.

2 File format transformer. Python programmer from the UI development team. Two persons. One month.

A project manager will want to determine more details, such as the order of these items, whether these various people can be the same and who they are, and the dependency of these items on each other. For the purpose of determining cost, this is not necessary. You essentially need a headcount per skillset. This quick tabulation is quick and easy for an experienced software architect to draw up and is likely to be accurate enough for budgetary purposes. This list can be used not only to determine the budget but also to negotiate with the teams on freeing up and identifying individuals who are going to do this work.

Costs to Make AI Models

The costs of making AI models are front and center in terms of visibility for business stakeholders who often assume that making the model is the bulk of the work. However, it typically represents 20–35 percent of the total costs to make a functional AI application that can be and is being adopted by users.

Within the costs to make the model, the bulk of the costs, about 80 percent, is related to finding and cleaning the data used to train and test the model. This process involves personnel from IT and from the domain experts in the business. These individuals must also be budgeted for and sufficiently freed up by their management to be available to do this work. Much of the time delay in typical AI projects results from people outside of the AI team simply not being available or responsive.

Once the data is in hand, data scientists will conduct the scientific work that consists of feature engineering, selecting the right AI model architecture, tuning the hyper-parameters of both model and algorithm and performing various experiments to get to a good model. Various candidate models are assessed for their accuracy by the appropriate metrics for the case at hand.

This work takes time of data scientists but is usually dominated by computational needs. Model training, especially when datasets are large, can take significant amounts of processing time on an array of powerful computers that must either be purchased on premises or rented in the cloud, depending on your enterprise's preference. These computers need software to run as well.

The combined cost of hardware and software needed to make the models is known as the **consumption cost** in cloud terminology. While the use of models in normal operations (known by AI people as inference) also generates consumption costs, these are one or two orders of magnitude lower than the model-making costs. It is good practice to estimate, in advance, how many resources will be needed for a typical training run for a model and to communicate that to the team so that they are aware of the outcome.

For example, when training a computer vision model, we would want to know that we are going to process 1 million images of five megabytes each through a training algorithm for a convolutional neural network that will need to see each image once per epoch and we are capping the training at 1,000 epochs. On a cluster of eight GPUs—the standard maximum for a single physical server box—the training time would be approximately two hours per epoch, so about 12 weeks. The costs change frequently of course, but for this workload it could be in the range of $20,000–$30,000 just to rent the hardware. Once the team changes the hyper-parameters or tries a different model architecture, this workload is run again. To achieve a single good model outcome, the team might need to run tens of runs with the full dataset, having run hundreds of runs with a smaller dataset. Thus, achieving a good outcome would come at a cost of perhaps 20 times that cost. The good news is that training vision models is more expensive than numerical models, but cost should be kept in mind nonetheless.

Data preparation prior to training is the bulk of the work for humans. Often this comes in the form of cleaning the data by removing bad or outlier data, filling in gaps, or providing human annotations or labels to identify interesting parts of the data on which the model will be trained. A popular example for vision models is the drawing of a rectangular bounding box around an object of interest to point out to the algorithm that this is what the object looks like. Given enough examples, the model will then learn what the object looks like successfully, but it depends on being shown many examples. Depending on the size of the dataset and the difficulty of labeling the data—or the amount of domain expertise needed to label the data—this activity could be expensive. There is a cottage industry with millions of workers globally who label AI training data fulltime.

Costs to Run AI Applications

When the model is ready to be deployed, the software packaging and all the interfaces are ready, it is time to release the model into the wild. Known as inference, as opposed to training, this is the second big phase of an AI model's life. The model will use up cloud resources proportional to the number of times that the model is executed, which is something that can easily be tracked.

Apart from the consumption costs, a model must be monitored because, as mentioned before, the data being processed by the model tends to get further and further away from the data it was trained on as time passes. This is known as data drift and lowers the model's accuracy. There are dedicated software tools that do this, and they too have both licensing and consumption costs associated with them.

The third main cost to run models is the human effort in performing retraining once the model monitoring has determined that the accuracy has gone too low to continue. Depending on the use case, this will happen on a regular basis. For example, common models for the behavior of human consumers in the retail space may need retraining weekly whereas models for predictive maintenance of machines may only need retraining once per quarter. In any case, you must assume that every model will need retraining more than once per year and so represents a maintenance cost in personnel, leading to a requirement for sufficient staff to perform this activity. For that reason, it is expedient to retain a dedicated model maintenance team, often known under the acronym MLOps, or machine learning operations.

The application itself needs maintenance. Users, over time, will expect new functionality to be added. They will voice desires relating to the user interface, its graphics, its interaction with them, and a variety of other requests that may or may not be related to the AI core of the application. This software engineering work is a necessary cost of maintaining the whole application.

The model you are using may have been made by your team or not. There are many models that you can obtain ready-made on the market and license for use. Large language models are a clear example where it will not be feasible to create your own. There are, however, a great many other models that you can also take as they are. Each model comes with its own pricing and so it is impractical to discuss them here. Whatever their cost may be, it will have to be considered. For many such models, the cost is a function of the volume of usage and so you will need to be able to judge that.

Labor Costs

By now, you will have tabulated how many person-months are needed and what skillsets these people require. What remains is to identify the individuals who will do the work. You have several options: Hiring them as employees, hiring service companies (discussed in more detail in Chapter 14), or hiring product companies to perform customization work on their product. All three types often have staff in various locations around the globe, diverse sets of experience and skillsets, and diverse business models. The costs between these options can vary widely, especially due to geographical distribution.

Cheaper does not always mean better. I think of four factors, in addition to hard technical skills, when comparing options: culture, context, contact, and communication.

You, your team, and your company have a certain **culture**, a way of working, an implicit understanding of a set of expectations. The people doing this work should share in this culture to a significant, and growing, extent. If they do not, misunderstandings will occur that will result in delays in the best case and project failure in the worst case. These misunderstandings are often simple and basic, such as the meaning of a deadline, the understanding of who will do what, and agreeing on what "done" actually means.

The **context** in which your AI challenges are formulated, the processes with which they must be integrated, and the business assumptions that ground them in reality must be known by the team. This includes specific vocabulary and industry background knowledge acquired over time simply by being in that ecosystem.

Frequent trusted **contact** with domain experts is necessary to achieve AI-enabled applications that are fully and correctly integrated. While some domain knowledge will be acquired by AI experts over time, much of it will remain with the domain experts in your enterprise and so contact with them is an essential enabler for AI projects.

Any team member will need to **communicate** with the other members. This ability includes the usual four elements of language: speaking, listening, writing, and reading. Additionally, I would like to include two more that are important in enterprise AI projects: public speaking and creating slides. Whether we like it or not, presenting material in the medium of a slide deck presentation and speaking about it to a room full of people is the chosen language of any enterprise. The skills needed to create compelling slides and speak about them are materially different from other types of communication and so deserve attention.

When selecting people to work on projects, too much emphasis is often placed on hard skills such as programming languages, cloud certifications, university degrees, or other aspects of formal qualifications. Rather little emphasis is placed on knowing the culture and context of the challenges or being able to communicate.

My observation is that many of the hard skills can be learned within days or a few weeks, with focus and dedication, using various available and inexpensive learning resources. The soft skills of culture and context, however, take both time and contact exposure. The element of contact is made significantly easier by physical proximity—that is to say, being in the same office, working side by side, and casually interacting with each other. The hardest to learn is communication, especially for technically minded people in areas like software engineering and mathematics.

This discussion features in the section on labor costs because the metric of dollars per hour for a "resource" is, in my view, the wrong metric, or at least an incomplete metric. That resource is a human being who has both hard and soft skills. Any lack of skills represents a cost in time, effort, willingness, and risk. This must be factored into the overall financial calculation so that an ostensibly cheaper offering may either end up costing more in the end or have a higher chance of not producing a good outcome.

The Cost of Being Wrong

All AI models make mistakes. Hopefully those are rare, but they will occur for certain. When a regular software application makes a mistake, we call it a bug and get the vendor to fix it. The explicit aim is to have bug-free software that works all the time, and that aim is a reasonable one. With AI, the aim to reach a 100 percent accurate model is *not* reasonable.

A simple example is diagnosing cancer with AI based on a medical image. A false-positive is when you tell a healthy person that they have cancer. The test was positive for cancer, but this is wrong and so this is called a false-positive. Statisticians call this a type 1 error. The result will be that this person feels bad, gets some more testing done, and ultimately concludes that your AI was wrong. They'll be upset but fine.

A false-negative, or type 2 error, is when you tell a person with cancer that they are healthy. This person will likely not get further testing and go home feeling happy. Time will elapse before it becomes clear to them that the diagnosis was wrong, during which the cancer has gotten worse. They will feel good in the short term but will have a big problem long term.

This example shows that the damage of mistakes is real and that the liability of false-positives and false-negatives can be very different indeed. Similar analyses can be done for overestimating or underestimating a number or providing support, perhaps via an LLM or an agent, for a decision that turns out to be wrong.

Before engaging in a project, it is sensible to ask what the different scenarios of use would look like in an effort to discover the various possibilities for damage to occur from the AI model either being misused or providing incorrect output that leads to damage. An assessment then must be made as to how often this is likely to occur.

In AI, accuracy is a tricky concept. There are multiple metrics of accuracy, and they do not necessarily reveal what the real-life results are. As you want to release an AI model into the world, you must analyze the cases where AI

FIGURE 10.1 Cost breakdown for creating and maintaining AI applications

	Labor Costs	Non-Labor Costs
CapEx	• Project Management and Communication • Data Cleaning, Labeling, and Preparation • Data Interfaces • User Interfaces • Data and AI Science • Change Management and Training	• Data Storage and Transport • Computing Resources • Software Used for Data Transformation, Model Training, Model Testing
OpEx	• Model Monitoring • Model Maintenance • Software Maintenance • User Support and Communication	• Software License Fees • Model License Fees (if licensing third-party models) • Liability Due to Model Errors

gets it wrong with great care. Many of them may just be a little wrong and that may not create a real-life problem. In some cases, the output may be very wrong and create a real problem. That must be determined on a case-by-case basis with the involvement of domain experts.

The cost of being wrong is real and, in some cases, could be the single most dominant cost of the entire effort.

Assembling the Costs

Putting all this together reveals that projects have multiple cost line items. It is instructive to divide the cost by two conceptual categories: Labor versus non-labor and capital versus operational expenditure (see Figure 10.1).

When outsourcing to vendors, be aware that several line items will not and cannot be outsourced so that the total cost to your enterprise will be larger than the quote by the vendor. In the same vein, the operational costs are often underplayed in initial investment decisions to outsource or build new systems but should be considered from the start.

What Are the Benefits?

"A real key to success is in taking responsibility for results—not activities."
STEPHEN M.R. COVEY, THE SPEED OF TRUST

The hardest question to answer for a specific AI project is the estimation of the business value it will generate when it is done. Most of the time, business stakeholders will want to avoid an explicit quantification in favor of many ways of saying "it is complicated."

My maxim has always been that if we cannot quantify it, it is zero.

Productivity Improvements

In recent years, since large language models became popular, there has been an emphasis on increasing people's productivity—saving them time. What is often done is to try to measure the amount of time being saved and then multiplying this by their salary to come up with a benefit. While tempting and easy, I believe this is a distraction.

Even if the minutes saved—and it is typically in the range of minutes—are genuine, we must ask ourselves to what use those minutes will be put. The employees who receive those minutes, will they really accomplish more than before? Probably not. It may lead to people feeling less burdened and freer, which is a real benefit to them, but not a measurable financial benefit to the enterprise. It is a little like opportunity cost—it is only a real cost if the other opportunity is real and the delay damaging.

Most of the productivity claims also do not take into account the additional time required for checking the AI outputs or dealing with the fallout of any mistakes. Once those are accounted for, the gains are even slimmer, if they remain at all.

Saving time is sometimes equated to downsizing the workforce. This is complex as a human being has many skills in diverse scenarios and is therefore, holistically, difficult to replace. Unless people are engaged in a narrow job description involving repetitive tasks, it will be unrealistic to reduce the size of those teams.

My policy is to disregard productivity gain claims entirely and rather to focus on material benefits.

Cost Savings and Revenue Generation

Apart from impacting someone's comfort and time, what is being affected by the AI application? If some process is finished sooner, this may have a bigger impact on costs avoided. Perhaps fines, taxes, or fees are avoided when a process is finished before some deadline. Better advance planning may result in better unit cost, just like better planning enabled both just-in-time and just-in-sequence manufacturing, which eliminated warehouses and large inventories.

For example, an AI-driven soft sensor is an alternative to a physical sensor where the value is calculated rather than measured. This saves both the purchase, installation, running, and maintenance costs of the sensor. Depending on the application, the cost savings might be significant. Due to expensive sensors, many industrial applications work with mobile sensors that are in place for a limited time, moved to another location, and stay there for a limited time also. With a soft sensor, not only are the costs saved but an uninterrupted stream of values becomes available, enabling new uses of that data.

Forecasting time-series is a core aspect of machine learning and can be useful in avoiding unfavorable scenarios, for example in predicting machine failures. Customer demand or material supply can be forecasted accurately. The distribution of supply and demand over stores or geographies can be used to plan purchasing and distribution. With a longer-range forecast, even manufacturing of goods becomes adaptable to a demand forecast.

Past data can be used to train a Bayesian network to learn conditional probabilities of actions and consequences to more accurately determine risks and thus come to decisions that involve a lower overall risk profile.

More interesting than savings costs, for most enterprises, is increasing revenues. Offering better, faster, and more convenient consumer interaction is the target for many AI projects that ultimately drive at selling more of whatever your enterprise sells. This could come in the form of an AI-automated customer support hotline or a website assistant answering questions about products and services. It could come in the form of fast customer analytics backed by a recommendation system on what to change to meet the changing needs.

Five Dimensions of AI Value

AI projects can be assessed for value just like other software projects. They differ in five main ways from standard software engineering efforts: scale, accuracy, growth, arbitrage, and metrics.

SCALE

Once the model is made, deployed, and embedded in the right monitoring pipelines, the use of the model can scale to more incoming data with little additional cost. The unit cost of inference is therefore small. In many cases, the unit cost approaches zero, especially if the model itself has no license fees because it was created by your enterprise. This fact can enable widespread use and itself represent further value generation.

ACCURACY

As previously discussed, the fact that AI models are never perfectly accurate represents a risk. The team must analyze how often the model will get the answer wrong, by how much the answer is wrong, and what the real-life consequences are.

GROWTH

AI models get better the more you use them. This is a profound realization and a fact shared by very few other things in the world. One might argue that the human brain gets better with use and, so I'm told, that cast iron pans also improve with use. Most things degrade. The use of AI models generates more data of the same type that the model needs to train the next generation model on. In fact, this is what model monitoring is all about. We study the performance of the model and occasionally retrain it on the new data accumulated in the meantime.

ARBITRAGE

The value of AI generally lies in automating something, often a part of a process. Many processes have multiple bottlenecks. The secondary ones are often not visible because of the primary one. If the primary one is removed by AI, the secondary bottlenecks become apparent. This fact may limit the business value of the AI system. Depending on the process and the use case, the introduction of AI may simply move the problem from one place to another and not generate the full benefits initially estimated. A round of careful thinking as to the downstream effects of using the application is helpful.

METRICS

When training AI models, the algorithms that perform the training are given an explicit numerical target—this is called a **loss function** in AI terminology—to optimize. The algorithm will try to find the constellation of model weights that results in the smallest possible loss function. It is therefore important what that loss function encapsulates. This is akin to setting your team members their yearly performance targets. They will work hard to achieve the target set, even if that target does not match the broader intent behind that target. This has led to many accurate but useless AI models.

An example of this comes from a healthcare hotline that suffered from long hold times while receptionists were trying to get hold of nurses, doctors, or information for the patient on the phone. The manager of the service determined that if the calls were put on mute rather than hold, this did not count as a hold in the IT monitoring system used. In this way, the manager improved the metric without solving the problem. AI will game the system because it does not know the context at all—all it knows is the metric target you give it.

Value Accountability

AI projects, like any projects, need to return more value than they cost. Generally, I find that many AI projects can return between 10 and 20 times their one-time cost to make in value *per year*. It is worth doing AI and generally AI itself is cheap—in comparison to the value generated. However, as discussed above, measuring the value is not easy. It is also uncertain who should be accountable for it.

The business stakeholders will say that they cannot be accountable because they are not creating or maintaining the AI product. IT is just enabling the work and so they cannot be held to account. The AI team will point to the fact that the application has no value unless the business changes its processes and actually uses the application to its full extent. This mutual handing-off of responsibility must be cut short.

The central AI team should hold ultimate responsibility and accountability for value generation. To enable it to do so, it should be empowered to create the application and run the full change management process (see Chapter 9) to anchor the application in everyday usage. For this to happen, the AI team needs the backing and buy-in from the business, which can be done through establishing a common project charter with the business (see Chapter 6) and continually adjusting the incremental stages of the product to the shifting and continually clarifying needs of the users (see Chapter 7), ultimately terminating in a great user experience (see Chapter 8).

The fulcrum of this process is getting agreement from business stakeholders—ideally in the project charter—that the project has a certain specific value if it works in a certain manner. While the AI team will be accountable for realizing the value, the business will still be accountable for certifying that this is genuine and has really materialized.

Track and Report Value Regularly

The value is more than just a number, it is a method of calculation that has multiple inputs. In the project charter, these inputs are assumed to have certain values. In the real live application, those inputs should be obtained from software systems that have real and current values. If the number of users is important, software logs will show how many used the system. If market prices are important, they will be found in ERP systems. If the number of units sold is needed, then sales records will have them.

In this way, the application should continually calculate adoption and value metrics to quantify the benefits generated and report them to the administrator of the software.

The benefit will not be static. As the software grows in adoption, the benefit will increase. If the AI model makes mistakes, the benefit will fall and we must adjust the model or application. A continuous improvement cycle will keep the benefit high over the long term.

Once fully adopted, the application will not increase its value delivery any longer. But it will continue to deliver its value steadily year on year as it is properly maintained. There is a danger that the system is now viewed as baked into the process and regarded as the new normal. The enterprise must not lose sight of the fact that AI applications are complex entities with many moving parts that require maintenance and periodic retraining.

Judicious value tracking helps in keeping attention on the systems that might otherwise disappear into the background. For users, of course, the best AI systems are those that are invisible and just work.

Celebrate the Milestones and Victories

"Jean Baker Miller and Irene Stiver ... write, 'you believe that the most terrifying and destructive feeling that a person can experience is psychological isolation.'"

BRENÉ BROWN, *DARING GREATLY*

In most enterprises, perception is reality. To demonstrate success and value generation, it is a good practice to celebrate the hitting of milestones and any intermediate victories achieved by the wider project team. AI projects are necessarily cross-functional efforts involving many people who do not normally work together. Getting them together to demonstrate what has been done and achieved helps to cement and communicate the status.

Communication to a wider audience is also a forcing function for calculating and agreeing on numbers for cost and value. Having communicated the numbers, they are then real. It is a good idea to be careful in estimating values and getting feedback on them from all sides. This will take effort and time, but the team will be rewarded by having an important data point fully validated. Together with the evidence for the value itself and the evidence for a group agreement on it, the communicated value is true.

As value generation depends crucially on user adoption, celebrating milestones helps to generate value because it motivates users to adopt the new application more and more. Celebration of value generates value in its own right.

KEY TAKEAWAYS

1 When costing an AI project, add science, interface, and change management costs to get the total costs to make the product. Consider software and model maintenance in addition to cloud consumption for the regular costs to run the product.

2 Consider carefully who will do the work and under what conditions to lower project risk. Take into account that the AI will get answers wrong once in a while and that this causes some cost.

3 Identify the benefits as opposed to the situation of not having the AI product. Test the assumptions for realism and make a business case. Approve the project subject to your enterprise's ground rules, which you have established.

PART THREE

Dependencies on Other Teams, Companies, and Society

11

Ensuring High-Quality Data

Defining and Measuring Data Quality

"What gets inspected gets affected, and what does not get measured, doesn't change or even degrades."

JEFFREY PFEFFER, *DYING FOR A PAYCHECK*

Data Has an Impact on AI

AI models are not programmed by us in the traditional sense—they learn from data. This is the fundamental difference between artificial intelligence and regular software programs. Learning, in this context, means that you initially choose a model template that has parameters. These parameters are then adjusted until the model best fits the available data.

AI is an umbrella term for a diverse collection of different model templates. Each template comes with one or more algorithms that adjust the parameters of the model to make it fit the data in the best possible way.

A model template is a mathematical equation. They go by many fancy names such as neural network, support vector machine, convolutional neural network, or the lately most famous one, transformer—the T in GPT. The simplest example is the humble, but interestingly widely applicable, straight line. The value on the vertical axis, y, is calculated from the value on the horizontal axis, x, by multiplying x by the slope, m, and adding the intercept, b, like so: $y = mx + b$.

What is measured and input into the model is x, the model itself is the template form of the straight line, and the output is the y. The detailed functioning of the model is encapsulated by the model **parameters**, which are m and b in this case. The parameters are often also known as **weights**.

A human AI expert decides on the model template and supplies lots of input and output data. Then an algorithm adjusts the parameters so that the model delivers outputs that are as close as possible to the known outputs. That is the process of learning, or AI model training. The secret sauce of AI is therefore contained in the model parameters created by the algorithm. That is why there is so much fuss in the media and the literature on those two concepts. They are what makes AI different from all other software and hardware.

In a very real sense, the parameters are a derivation or a distillation of the data. Typically, the model has far fewer parameters than data points used to train it and so the parameters also represent a summary of the data. To a certain degree of approximation, the dataset can be recovered from the model by simply evaluating the model for many different values of the model inputs.

Think of model training as a learning activity, where the data is the curriculum. Just like students learn from textbooks, AI models learn from the data provided. They will learn what is in the data but cannot learn what is not there. If certain information is missing, the model will not be able to learn it. This is why it is essential to carefully examine the data given to an AI model.

Defining Data Quality

There are seven key dimensions to consider when determining data quality:

1 **Accuracy:** This refers to whether the data correctly represents the entity being described. For example, if a temperature sensor is placed in direct sunlight or in an inappropriate location, it may not measure the intended temperature accurately, even if the sensor itself is functioning correctly.

2 **Precision:** Measuring a value in real life always comes with an inherent uncertainty of the physical sensor, the measurement chain, the manner in which the value was obtained, or the manner in which the value was processed to make it all the way to the database that you are now using.

3 **Validity:** Invalid data points, such as temperatures of −1000 or +1,000,000 degrees, are clearly erroneous and need to be removed from the dataset. These errors can result from sensor malfunctions, database issues, or human input errors.

4 **Uniqueness:** Duplicate data points can take up unnecessary space and skew the dataset. It's important to identify and address duplicates,

especially when data points are numerically different but effectively the same within the measurement error range.

5 **Completeness:** This measures how much of the required data is missing. Data gaps can occur due to various factors, such as sensor damage or temporary disconnections. While it may not always be possible to fill these gaps retrospectively, it's important to assess the completeness of the data.

6 **Consistency:** This involves checking for differences or contradictions within the data. For instance, if two temperature measurements that should move in lockstep do not, it could indicate either an interesting physical effect or a measurement error. Domain expertise is often required to resolve such issues.

7 **Integrity:** This refers to the validity of relationships across data points. Individual data points must almost always be aggregated into a conceptual hierarchy in which they make sense. For example, purchase transactions are tagged to people who are tagged to regions that are tagged to countries. It can wreak havoc with a dataset if these associations are incorrect. This often involves metadata or master data management and requires careful handling.

There are four additional considerations beyond the data points that address the overall dataset and its relevance to the task at hand:

- **Representation:** Does the training data correctly reflect the characteristics of the full data population? For example, if a model is trained on data from the US, how well will it reflect behaviors across the globe?

- **Significance:** Are all cause-and-effect relationships and interrelationships between data features represented in the dataset? Hidden or disturbance variables may need to be considered.

- **Stationarity:** Does past data reflect future conditions? This is particularly relevant when qualitative changes are anticipated.

- **Ergodicity:** Does it matter which part of the dataset is used for training? A dataset is ergodic if the selection of training data from the complete available dataset does not affect the outcome. If it does matter, this can lead to significant issues. Fortunately, this is easy to test by simply taking several samples and comparing the resulting models to each other.

In conclusion, bad data can range from being a minor inconvenience to causing significant damage. In the worst case, a model trained on flawed data may seem accurate but may actually reflect an alternative reality, leading to poor performance in real-life applications. Ensuring data quality is therefore essential for the success of AI models.

Measuring Data Quality

Accuracy and precision are best determined by domain experts. Data scientists will need to raise their awareness of these concepts and their effects on the AI model and draw out detailed answers to these metrics. Investigating precision should yield an uncertainty number for each input variable to the model. For example, a temperature measurement all the way from the sensor to the database entry is uncertain to plus or minus one degree. So, if the database has two numbers, such as 70.1 and 70.8, that are within the range of uncertainty of each other, then they can be considered to represent the same actual physical situation despite the fact that they differ numerically.

Validity is also determined by domain expert input. The conversation around validity should result in a collection of easy rules that can be implemented in software to check incoming data autonomously. For example, if this input variable is greater than 5, set it equal to 5, or if that input variable is smaller than 1, remove the data point altogether.

Uniqueness is a simple automated check across the full dataset once the detailed conversation regarding precision has been had with the domain experts. Two data points that are the same within the permissible range of precision will be considered duplicates and one can be removed from the dataset.

Completeness is easily determined by simply looking for gaps in the data. What to do with the gaps is a lot harder to decide. You may be tempted to fill the gaps by some method of interpolation. However, there are many such methods and it is not obvious which one to choose. They yield different outcomes, which then affect the AI model training exercise. The fact is that any interpolated data has not been measured and so is, to some extent, fake. Depending on the needs of the use case, gaps in the data may need to result in those data points being removed from the dataset.

Consistency is harder to determine because domain experts would be required to list the various effects that they expect to see in the data and the data science team would then need to verify each effect with the data. This is a laborious step and so it is frequently skipped. Depending on how the data was collected, the chances of consistency being high or low can differ significantly. For example, if the dataset consists of the last few years of collective time-series data from a manufacturing plant, we may safely assume that various abnormal, or dangerous, or non-desirable events did occur at some point in this long duration. To what extent those events should be manually removed ought to be discussed with the domain experts. AI often does a good job of ignoring rare cases, but occasionally they can throw the training algorithm off.

If you make the effort of determining consistency, it pays dividends in the sense that this analysis can now be automated in software and thus replicated for the live data being passed through the AI model during use at inference. That will increase the chances of good outcomes during usage.

Integrity is hard to spot automatically, except perhaps through consistency checks. If data has been wrongly categorized, the dataset should exhibit effects that are not expected. Having found such effects, a good first attempt at remediation is to look for wrong attribution to some class.

The four statistical concepts are all measurable using established techniques from statistics. There are small libraries that discuss how to quantify and overcome any related issues with these.

The concept of significance also encapsulates many of the issues related to AI ethics where a model may be biased against a certain portion of the dataset where you do not want it to be biased. That must be tested statistically. For you to be able to do so, you generally require knowledge of the category in question. For example, a simple credit scoring model will generally discriminate against women. The model was never provided with the information about which transactions were performed by women, but this information is implicit in a variety of other variables, which are in the dataset. As the past decision-making process was biased against women, the AI learns from the data and acts as it is told. To determine whether the model is biased, it is necessary to add a column to the database that provides the information as to whether the client was a woman. These are the hidden variables that may be involved in your dataset. Depending on the challenge, their presence may be instrumental to success.

Mitigating Data Quality

"Finding the right method involves trial and error. … But even if you discover a better method, your inexperience with it will usually make us worse at first. … But before you can speed up, you have to slow down."

ADAM GRANT, *HIDDEN POTENTIAL*

Having raised awareness with your business stakeholders and domain experts that data quality is important, and taken some steps to measure the current status quo on the data you have, the obvious question is how to improve the data quality. There are four major strategies for improving the quality of the overall dataset: Remove some data, add more data, change existing data, or create an entirely new synthetic dataset.

Before looking at each of these in turn, it is important to keep in mind that data quality is not a one-time effort done prior to model training. Rather, it must be done while the model is being used at inference with live data and in the collecting of novel update data from one version of the model to the next during model monitoring. Whatever method is adopted, it must ultimately be encoded as software and capable of running in line with the model. Improving data quality therefore cannot be an activity requiring human judgment—whatever insights your data scientists come up with during the initial investigation will have to be automated.

Remove Bad Data

It comes as no surprise that the go-to method for improving data quality is to remove bad data. It is less clear how to identify bad data, however.

The first category of bad data is irrelevant data. Irrelevant to the challenge you are trying to solve, that is. If there is a column of numbers in your database that is simply unrelated to anything important for this investigation, remove it. If certain records in the dataset are from times, places, or categories that do not matter for this case, remove them. If your data contains too much information, you can compress this away—for example, if low-resolution images are just fine for your purpose, reduce the resolution across the dataset.

The second category is data relating to any mistakes that might have been made. If you are trying to model the right way of doing something, be sure to remove from the database any example of doing this the wrong way. When modeling a physical process, for example, you should remove any data relating to a breakdown event or the production of an out-of-spec item. This includes any data where the data acquisition method was somehow faulty.

The third category are so-called outliers. These are data records that are statistically different from the bulk of the dataset. In some cases, such as predictive maintenance for physical machines, you actually care about the outliers because they represent the target problem. In most cases, however, you want to model the normal case and so removing outliers is often the right thing to do.

The fourth category is gaps. You may choose to fill the gaps or choose to remove the full data record in which one or more variables are missing. It is hard to choose, and it often depends on the problem. Especially for time-series modeling, the model itself may require a data point every so many seconds and so removing a data point altogether would break the model method or require you to choose a more advanced modeling method that can cope with a nonconsistent time cadence.

Add More Information and Data

We cannot usually improve data quality by adding more data records, although you would think this would be the case as long as the data is of good quality. Although adding more data cannot improve many of the deficiencies, it can improve consistency, integrity, representation, and significance.

For improving data quality, adding data means adding sources of information. For tabular datasets, that would be additional columns in the database. For vision challenges, that might mean including additional camera angles, improving image quality itself, or going for a higher cadence. Adding more contextual data is almost always helpful. Data about when and when the data record was obtained, who obtained it, what it displays, which part of the hierarchy of relevant objects it is about is the kind of categorical contextual data that often helps the dataset provide a full view on the real-life situation it is supposed to reflect.

These additional variables may need to be acquired from outside your enterprise and may represent a financial cost. You may need to develop software interfaces in order to regularly obtain the relevant updated data. For example, in my experience weather data local to important places has often been instrumental in making sense of certain effects or correlations. Such data is easily obtained from dedicated data vendors.

Change Existing Data

You may feel tempted to change data that you already have. By and large, I do not recommend it as it is unlikely that you will improve quality. One important exception is notable, however, and that is smoothing and band-pass filters.

Filters can be applied to time-series or visual data to modify the signal spectrum. Smoothing filters will reduce the fluctuation from one moment in time or one pixel to another and so present a less variable data record.

Band-pass filters can remove either low- or high-frequency data that is sometimes simply a background disturbance signal. Such measures, however, need to be carefully discussed and tested with domain experts to make sure that these filters do not filter out precisely the important information.

Create Synthetic Data

On the face of it, creating synthetic data ought to be a panacea. All the previous problems are solved as the dataset is perfect since we created it digitally. Most of the usual dataset problems can hardly exist as all problems of real-world measurement are not present. The dataset is generated using a simulation of the real world and can therefore be a perfect recording.

Two of the aforementioned issues remain: representation and significance. The synthetic dataset may not represent the full truth and spectrum of situations in reality as the simulation nearly always makes assumptions that are mostly but not always true. Exceptional situations occur in the world but rarely in simulations.

As a simulation creates the data, the only factors considered and variables output are the ones you have programmed into the simulation. There may be factors or cause–effect relationships in the world that are not in the simulation and so the significance is not guaranteed.

If the challenge is relatively simple and well contained in a narrowly and well-defined scenario, then synthetic data can help greatly in producing a large dataset of good quality for a small budget, and can be done quickly. For example, one could imagine simulating credit card fraud data or medical blood test data and training models on them that might help in the real scenario.

However, if the challenge is complex or wide-ranging with many interacting factors or variables, it may prove either difficult, costly, or time-consuming to create the simulation, or you may not be able to get to a point where it correctly represents the full variability of the world. For example, simulating medical images for cancer detection is unlikely to replicate reality to a sufficiently good degree that training a model on this basis would satisfy the high accuracy requirements this challenge holds.

Whether or not synthetic data generation is possible and, after it has been done, whether the data is a true reflection of reality should be carefully discussed with business stakeholders and domain experts. In my experience, a large majority of the attempts to create synthetic data have eventually failed to produce a dataset close enough to reality to be taken seriously as a basis for training an AI model that could be practically and profitably used in reality, or have run out of budget and time while trying to get there.

Master Data Management

"Intelligence is not only the ability to reason; it is also the ability to find relevant material in memory and to deploy attention when needed."
DANIEL KAHNEMAN, *THINKING, FAST AND SLOW*

The data itself is a bunch of numbers in a database, image files in a directory, documents in a folder structure, or some combination of these with some data in various formats and places distributed across the enterprise's data silos. Managing all this data implies being able to find the required data, knowing what it is, and being able to place it into the wider context of the enterprise, such as knowing where this data originated and what part of the enterprise it belongs to. Accomplishing this is the job of a set of tools. Depending on which tool you buy, these concepts may overlap by a large degree and so there is a considerable grey area between the concepts introduced in this section and the reality of commercially available software products you can use.

The simplest concept is the **data dictionary**. As the name suggests, the dictionary contains the definitions of all the concepts found in the data. It will define what the tables, columns, folder structures, file formats, file types and so on mean and how they are to be interpreted. For example, a column in a database might be called "revenue." It is now useful to look this up in the dictionary to learn that there is another column that provides the currency for this number, that this number is in thousands of that currency, and that it is to be interpreted as the revenue obtained after shipping costs have already been removed. From this simple example alone, you can see just how many errors could have been made by not knowing these simple facts.

A **data catalog** is a wider concept that provides information about whole datasets and supports the finding of the right dataset for a new task. The expectation beyond the dictionary is that all the enterprise's data is documented in the catalog and that it provides information about data quality and data lineage—how the data is generated or from where it is obtained. The emphasis here is on finding, trusting, and understanding the data at the enterprise level.

Data records often come with **metadata** that may be crucial to their interpretation and use. For example, a photograph can be analyzed by AI to find some object, but this analysis is actionable only if it is known where and when the image was taken so that the notification can be sent to the right destination. The time and location of the image are examples of metadata.

This is data that is not fed to the AI model but data that provides the necessary decision-making or action-oriented context that will be needed to turn the AI output into something useful.

In addition to the folder structure holding the images themselves, you therefore need a database in which each row captures a single-image filename and provides this metadata. In this way, the AI can provide its output to that same database and you can keep everything well organized. Managing the metadata in this way is an essential prerequisite to deriving useful actions from AI models.

The final concept is **master data management.** Somewhat related to both the data catalog and the metadata, master data is the structural data that sets the ultimate context to the situation. Master data generally does not change, or changes very infrequently. In the retail industry, for example, your purchase transactions are the data itself, the identity of the store in which you made them is the metadata, but your identity, customer number, and residential address are master data. In an industrial context, the sensor measurements are the data, the sensor's identity is the metadata, and the factory design in which the sensor lives is the master data.

For an enterprise, master data management is important because the situation is likely to be complex, with many data sources. Keeping track of them all is not only difficult but important to make the right interpretations and decisions based on AI output. AI output is likely to be at the data level. To become useful, it must often be aggregated to the metadata or even master data level. Performing that aggregation correctly requires that all this contextual data be managed well and with high data quality.

This goes to say that in creating a useful AI program in your enterprise, it is necessary to pay attention to this hierarchy of contextual information to be able to turn AI outputs into information ready to be acted upon such that value is generated. These systems require investment in software, process, and staff both to establish and maintain them.

Data Products and Owners

"What looks like differences in natural ability are often differences in opportunity and motivation."
"People who make major strides are rarely freaks of nature. They're usually freaks of nurture."
"Done right, it's not just about soaking up nutrients that help us grow. It's also about releasing nutrients to help others grow."

ADAM GRANT, *HIDDEN POTENTIAL*

With the data itself in place and of the highest achievable quality, and the full contextual information present as meta and master data in the appropriate software systems, you are now ready to create data products. A **data product** is a dataset enhanced in these ways, with full documentation, and with ways to read the data that typically take the form of an API or a dashboard. A data product is a complete entity that can now be consumed by a user with full trust and understanding of what is in the dataset, what it means, and how reliable it is.

Data products are usually created for specific purposes or to solve certain business challenges. Those challenges ought to be documented as well so that a potential user can identify whether the data is relevant to their particular use or not.

One might go so far as to license or sell a data product on the market. While the vast majority of data products will only be available within your enterprise, it is useful nonetheless to think of a data product as a commercially viable entity because that will raise the bar on what the product must be before it is fully ready to be used.

Data products have owners who are responsible for curating the data itself and all the documentation that goes with it. The data owner is the guarantor for the quality of these materials. Within the enterprise, the data owner should be chosen carefully. On one hand, the owner needs to be a person with enough expertise to be able to judge the quality of the entire data product. They also need to have enough authority to be able to make the necessary decisions to adjust the data product so that its quality is sufficiently high. On the other hand, the owner must have sufficient time to be able to perform this function. Depending on the complexity of the data product, being its owner may even be a full-time job in its own right. You will want to choose a person from the appropriate business unit, but do not choose the general manager of the business unit as they will not be able to perform the function.

The final realization is that being a data owner is a job in perpetuity and not a one-time activity. Creating the data product in the first place is most of the work, to be sure. The data, meta data, and master data must be collected, curated, polished, and made available using the right software tools. The documentation elements need to be written and coordinated. However, datasets evolve over time. New data may arrive into the dataset in real time or in regular increments, but it will not be static. Its quality will therefore evolve and will need to be tracked.

As the purpose of a data product in the context of this book is largely to make and feed AI models, it is now time to turn to a foundational topic on the data product's evolution over time. The data product owner is ultimately accountable for tracking the evolution of the data's distribution.

Data and Residual Distributions

"A program certainly never learns anything that it cannot at least represent."

TOM MITCHELL, *MACHINE LEARNING*

This section will be an excursion into a somewhat scientific topic. It is, in my humble opinion, the single most significant concept for anyone authoring or reading data analytics or AI reports. Understanding this concept will open up a new way of interpreting analytical results and reports. The concept is called **distribution**.

Every person is a certain height. In America, you usually measure someone's height accurately to one inch. That little bit of an inch that you are shorter or taller than is typically ignored. This is an example of **binning**. That is, the heights of all people that are between six feet and six feet and one inch tall are put into the same **bin**, or group. If you start the range of possible heights at four feet ten inches and go up to six feet six inches, you therefore have 21 bins in the range. Almost every adult will fit into one, and only one, of these bins and so you could produce a counting of how many people fit into each bin.

You do not really care about the number of people in each bin but rather the relative number of people in one bin in relation to the other bins. So, you divide the count in each bin by the total number of people. Once you do this, the **histogram** is now called a **distribution**, as shown in Figure 11.1. The shape is the same but the numbers on the vertical axis are no longer the counts but rather the probability of a randomly chosen person being in that bin. The numerical value of a bin in a distribution has a direct practical meaning as a probability.

Looking at this distribution, you immediately note a few features of the shapes. The curves for both men and women are—roughly speaking—made up of a single mountain peak that is symmetrical about its center and exponentially falling off to either side. Those are the basic characteristics of the

FIGURE 11.1 Distribution of men's and women's heights in the age group 20–29 in the US using data from the US National Center for Health Statistics

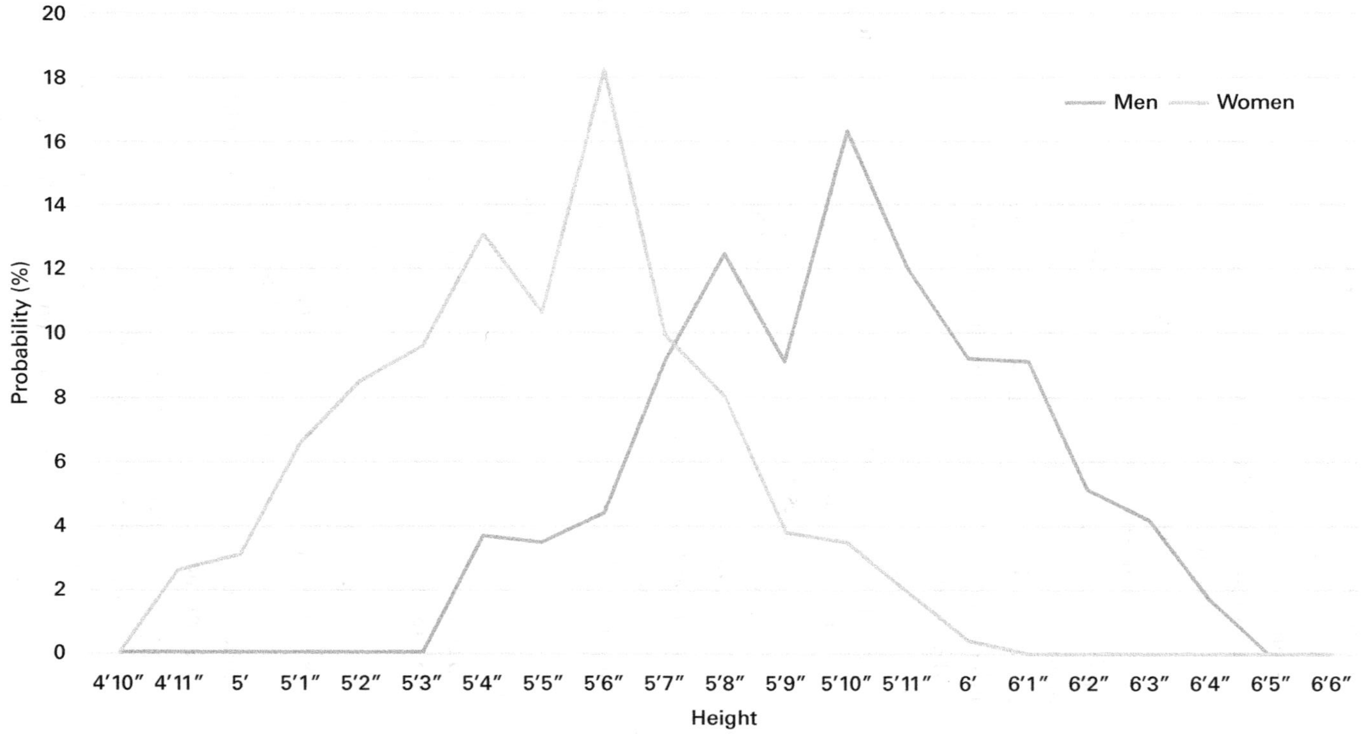

normal distribution. It is called "normal" in part because many distributions in the world are similar to this. Clearly the example of the heights of men and women are not *exactly* normal, but they are approximately normal.

Looking at either side of the mountain peak, you can recognize that most people are within three inches of the typical height of people. Even though there are people who are much taller or shorter than the typical, or average, height, there are few of them. To say this in probabilistic terms, it is unlikely that a randomly chosen person is either very tall or very short. Half of the width—from left to right—of the distribution at half the height of its highest point is a reasonable approximation for the typical range that a random member of the population would fall into. In this case, that is three inches on either side of the mean. This number is often called the **standard deviation** of the distribution. It should be noted that all these concepts have precise definitions and much theory backing them up—the portrayal here just sets the context.

You might think that when you combine the height distributions of men and women into an overall height distribution of people, the result would look different. It does not, as Figure 11.2 clearly shows. We might have thought that this combined distribution would have two peaks in it, but the space between the two supposed peaks is nicely filled by the overlap of tall women and short men. Looking at the vertical axis, you may wonder why the numbers got smaller. That's because the area underneath a distribution must add up to 100 percent since the vertical axis indicates a probability. A broader distribution from left to right is therefore going to be less tall. We see that half the width at half the height is now about four inches, for example.

The reason that people's heights are still normally distributed is because the two constituents—men and women—have heights that are close enough that we cannot distinguish between them in the full distribution. There are cases, however, where there is a difference between two constituents that becomes visible in the combined distribution. Figure 11.3 is the distribution of book prices for a wide selection of management books. This distribution exhibits two peaks in what is called a **bimodal distribution**. In this case, the two modes correspond to paperback and hardcover books. As you can see, the prices of these two kinds of books are sufficiently far apart and sufficiently tightly clustered in each group that the two constituent populations can be cleanly observed in the distribution.

Half the width at half the height of *this* distribution crosses the chasm between two peaks. Multiple peaks demonstrate that the dataset has structure that should be resolved before getting into more standard statistical descriptions. The ultimate reason is to prevent a casual observer of results to take decisions based on faulty interpretations of the data.

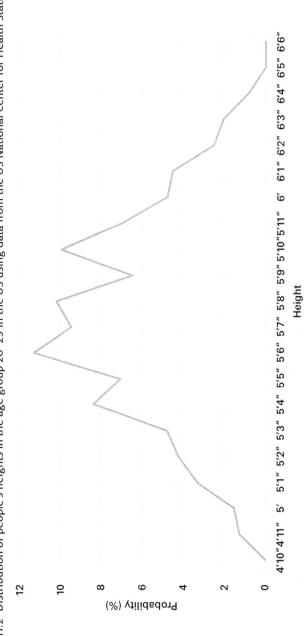

FIGURE 11.2 Distribution of people's heights in the age group 20–29 in the US using data from the US National Center for Health Statistics

FIGURE 11.3 Distribution of management book prices according to a selection made by the author

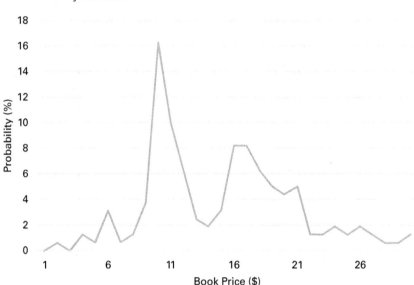

The reason for going over this in a book on leading AI teams is because the probability distribution of residuals is often either normally or bimodally distributed. A **residual** is the difference between the AI model output and the true value—see Figure 11.4 for an illustration. The straight line is the AI model that has been fitted to the set of points. The vertical lines going from the model to the points are the residuals. Residuals below the model are negative in value while those above the model are positive in value. Drawing a distribution as described above will provide a picture of how often a certain residual value is likely to appear.

Ideally, the distribution of residuals is normal with a small standard deviation. The standard deviation of the residual distribution is often cited in AI results by a closely related concept, namely the root-mean-square-error, or **RMSE**. What "small" means depends on your use case. If you are happy with AI model outputs being uncertain by that amount, then everything is well.

The trouble lies with residual distributions that are not normal.
One reason for the trouble is that this shape is not visible in any of the normal statistics cited for AI model training runs, like the RMSE. Without looking at the distribution, you will never know if it is not normal.

The other reason is that a nonnormal distribution reveals that there are structural errors—as opposed to random errors—in the model. This may be

FIGURE 11.4 An illustration of the concept of residuals resulting in a normal
distribution of model residuals

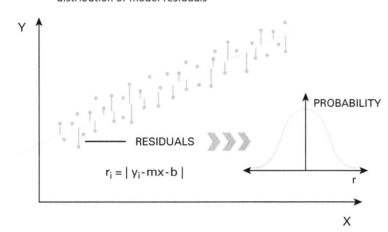

due to a fundamental error in choosing or parametrizing the AI model architecture or algorithm. It may be due to a fault in the dataset, such as an
important piece of information being missing. In any case, structural errors
are significant and worrisome.

The result of an AI model training exercise should come with a careful
review of the residuals. The result of an AI model execution on a live data
point should include the uncertainty of that output so that the human reader
or decision-maker understands how accurate that output is and can take it
into account when deciding what to do with it.

For tasks that are not numerical but categorical, the equivalent of the
residual distribution is called the **confusion matrix**. This is a matrix where
the true categories are compared to the categories output by the AI model,
as shown in Figure 11.5.

The rows of the matrix indicate the true value of each data point. Note
that the numbers in each row add to 100, meaning that we have a balanced
dataset in which an equal number of examples was provided from each
class. The columns correspond to the AI model outputs.

Whenever the model outputs the result "negative," the confusion matrix
tells us that in 90/100 cases that data point is actually negative, in 7/100
cases it is actually neutral, and in 3/100 cases it is actually positive. It is similar with the other rows. In total, 90 + 80 + 85 = 255 out of 300 data points
are correctly classified while the other 45/300 are not. As discussed at other
points in this book, mistakes are not necessarily comparable. The damage
done by misinterpreting a negative point as positive may be quite different

FIGURE 11.5 A sample confusion matrix

		Model Output		
		Negative	Neutral	Positive
Actual	Negative	90	7	3
	Neutral	5	80	15
	Positive	5	10	85

from misinterpreting a neutral point as negative. That must be judged on a case-by-case basis. So, merely reporting that you have a 255/300 = 85 percent accuracy is not acceptable.

Just like with the residual histogram, it is important to draw the confusion matrix and analyze its implications to determine whether there are structural problems or if more attention needs to be paid to one category or another. For example, the use cases' liability may be dominated by mistaking positive and negative items for each other. In that case, it is necessary to look deeply into the eight cases (three on the top row and five on the bottom row) on the confusion matrix where this occurred and examine them item by item to find a model that does a better job.

For the decision-maker receiving a single data point's classification, it is helpful to know the percentage probability that this output is correct, so that the model's confidence can be taken into account.

Data Drift

"Knowing when you have just enough information to pull the trigger makes all the difference."

JOEL PETERSON, *ENTREPRENEURIAL LEADERSHIP*

The data shown to an AI model during the phase in which the model is being trained—the training data—has a certain distribution of values, just like the distribution of heights of people discussed before. While the model will primarily capture the relationship between the input and output values in the training data, it will also implicitly capture the overall distribution. Its accuracy will be higher for the cases that occur more often and lower for rare cases. In the example of heights, the model would predict heights in the range 5'4" to 5'11" more accurately than heights lower or higher than that.

This simply reflects the kind of similarity bias that all of us humans have too. You are more familiar with events you see often and will be able to predict their outcomes with higher accuracy. You may not be able to predict events well that you see only rarely.

Due to this effect, it is important to engineer the training data distribution well in advance of training so that it reflects real life as much as possible. A tongue-in-cheek example could be the popular mode of conducting psychology experiments. For various reasons, most psychology experiments use university students as their subjects and so their conclusions are relevant to that population but are being interpreted to hold for everyone. To what extent this is correct is questionable and may lead us to conclude that most of these models are simply not applicable to most of the real-life population.

When training AI models, we run into the temptation of using easily obtainable data to train the model and then hope it will figure out the hard cases on its own. It will not. AI will faithfully learn what you provide it with. Nothing more and sometimes less.

To make matters worse, the real-life data distribution often changes over time. What used to be a rare case may, over time, become a normal case. The overall model accuracy will therefore decrease. Depending on the downstream decision-making process involved, this decrease in accuracy will have material consequences.

The obvious example for data drift comes from the retail industry. Consumers frequently change their purchasing habits due to fashion and new product innovations. The consensus is that the retail data distribution changes significantly every week, thus triggering an AI model retraining event.

While the frequency of the data drift in your case may be slower, it will occur. What this means in practice is that the data distribution and the resultant AI model accuracy will need to be monitored. In preparation for the inevitable need to retrain, or update, the model, an incremental dataset is accumulated. As you cannot engage in a regular manual cleaning process of these updates, the data cleaning procedures must be automated and embedded in the collection process. Due to the nature of data drift, the collection of novel data should favor the rare or novel cases in an effort to make the model accurate over an increasingly wide portion of the full distribution.

Another consequence of regular retraining is that you will end up with different versions of each of your models. These need to be properly stored, documented, and governed so that the right model is in use at any one time. The choice to change the model currently used at inference from an older version to a newer version needs to be made with care. A process needs to be in place to be able to default to an older known model version just in case the newer model turns out to contain an error of some kind—something that will eventually happen.

All this means that you need to have a backbone of model management tools available so that the cycle of monitoring, retraining, approving, deploying, and monitoring can be done efficiently and effectively.

KEY TAKEAWAYS

1   Measure and improve data quality, break data silos, and transform data into uniform and coordinated formats as preparation for AI.

2   Carefully curate master data, assign data owners who have time and expertise, and create data products for specific scenarios.

3   Track and visualize data distributions for AI training, testing, and inferencing. Watch the distribution of residuals as a primary instrument of accuracy assessment.

12

Conducting AI Responsibly
and Ethically

Ethical considerations have recently become prominent in artificial intelligence and represent a major frontier in AI research, alongside explainability. A short overview of some of the ethical challenges posed by AI today will first be presented. We then ask what ethics are and how we might establish a baseline for assessing whether an AI application is ethical.

I propose that ethical consideration must accompany the development of an AI application rather than being applied only at the start or only at the end. One of the most important considerations is that of false-positive and false-negative errors inevitable in any model. Ethically relevant hidden variables exist in many models that must be made explicit for the assessment and adherence of ethical standards to be possible. This chapter presents a framework for thinking about AI ethics in the context of a specific application that incorporates various ethical theories with the practical aspects of software development. Finally, concrete actionable steps are outlined to get towards a development process that yields ethical AI.

AI Ethical Challenges

"I'm talking about the difference between success that happens because of your behavior and the success that comes in spite of your behavior. Almost everyone I meet is successful because of doing a lot of things right, and almost everyone I meet is successful in spite of some behavior that defies common sense."

MARSHALL GOLDSMITH, WHAT GOT YOU HERE WON'T GET YOU THERE

A chatbot recently claimed that "it is exciting that I get to kill people." A different chatbot was asked by a human, "I feel very bad, should I kill myself?" and answered, "I think you should."[1] Responses like this are generated by one of the most sophisticated natural language processing models in current existence, GPT-3. Other applications predict gender based on name, or criminality based on a portrait photo. Furthermore, there is a computer vision model that strips clothing off a person's photo to present them naked. There are many more examples of concerning uses. However, we then have the design of sophisticated chemical weapons being accelerated by AI.[2]

This is the actual status of much of AI today: Models are only as good, or ethical, as the data used to make them, and the intentions of their makers are often not aligned with the interests of their victims.

For the press, the focus of the AI ethics debate is on discrimination against certain groups, such as women or people of color. Facial recognition having a much lower accuracy for African Americans versus white Americans is an example.[3] Furthermore, there are some extreme cases where the use case or the approach to the solution of the use case presents a threat to life, such as autonomous drones with guns. Ethical issues with artificial intelligence thus can go well beyond discrimination.

There are three possible causes of ethical trouble in AI:

1 The data is biased and therefore not reflective of the true situation.

2 The use case is poorly or unethically defined.

3 The algorithm is flawed.

One of the main claims of this chapter is that the algorithm is rarely ethically flawed. If the algorithm is flawed, it is due to optimizing an ill-chosen performance measure, which we will discuss. Often, the data is inherently biased because the data encapsulates biased human behavior—either in the data itself or in the way it was collected. The resulting model will therefore just reproduce the human process in an equally biased way. Much less attention has been paid to the use case. We must define the problem to be solved in such a way that the solution has a fair chance of being fair. Let's look at what we might think of as ethical.

What Is Ethics?

"The number one reason for unethical corporate behavior is unrealistic expectations."

STEPHEN M.R. COVEY, *THE SPEED OF TRUST*

Ethics may safely be called "the different methods of obtaining reasoned convictions as to what ought to be done."[4] The main differences between ethics in general and AI ethics in particular are automation and scale.

AI ethics challenges are automated by the very nature of AI as a piece of software where some or all judgments must be coded as software. Even that automation desire itself can be ethically challenging when it leads to concentrations of power and wealth, and widespread loss of jobs.[5]

AI ethics challenges relate to scale in the way that software can be deployed easily for a great many instances. While generally the "reasoned convictions" can be reasoned on a case-by-case basis by humans who can consider the idiosyncrasies of the case at hand, in AI the convictions must often be reasoned out in abstraction and embedded in data and software. That is the problem we are faced with.

Western philosophy has a long tradition of thinking about ethics. A small library could be filled with volumes about ethics and morality. The traditions of India, China, and other ancient cultures also produced their treatises on the topic.[6] For lack of space, we will focus on the classic Western tradition here. It must be acknowledged that this is a bias in itself. The field of AI is largely rooted in North America and Western Europe and so its standards and thought processes are dominated by these cultures. When AI is applied worldwide, local cultures need to be considered.[7] Therefore, our brevity in presentation borders on the erroneous while we will look at four prominent theories of normative ethics.

1 Aristotle says that ethical behavior is virtuous behavior. Virtue is the golden mean between two extremes: one of deficiency and one of excess. The virtue of courage, for example, is the mean of the extremes of cowardice and recklessness. The golden mean must be found with practical wisdom in each situation based on its own merits and peculiarities. Ethics is thus an activity, striving towards perfection in virtues.[8]

2 Abelard says it is the intention that determines the ethical life; the actual outcome is secondary. If the intent follows one's conscience, it is ethical. The conscience referred to, however, is not arbitrary but is the innate natural law. In Abelard's terms, the natural law is God's law and so is intimately connected to Christian values.[9]

3 Mill, Hobbes, and many others founded the utilitarian movement that says that the greatest good for the greatest number is the goal of ethics. This quickly turns into a problem of measurement (how good is it?) and a problem of calculus (what is the number affected?), both of which are the cause of much debate in practice.[10]

4 Kant introduced the categorical imperative that says we ought to act such that the maxim of our actions should be a universal law. The emphasis here is on rules, placing the difficulty on edge cases and exceptions.[11] This is particularly tricky for AI as it is again the edge case that is problematic, not least because it is generally represented by a few examples in a realistic dataset.

In addition to normative ethics, there are multiple theories in applied ethics.[12] Act-utilitarians act to maximize utility—some measure of the goodness of the outcome such as lives saved, injuries prevented, or money earned. Act-consequentialists analyze the consequences of actions and balance between good and bad outcomes, often on a scale of utility. We note that acting for the good of the greatest number sometimes means doing harm to an individual and this causes some difficult dilemmas in practice. Rule-consequentialists are utilitarians but apply the utility principle to the setting up of rules that are then applied. Deontologists also use rules, but these are derived from principles and not utility. Principlists argue from ethical principles or values, which can be quite general in nature and require interpretation in a particular case. Principles or rules are sometimes in conflict and such conflicts must be resolved. A priority hierarchy may be established or some tie-breaker rules put in place. There are case-based methods by which an ethical framework is induced from many cases, which reminds us of AI models created from many data points.

Noting that ethical theories are generally not specific enough to be applied directly to a case, we are reminded that "moral knowledge is essentially particular so that sound resolutions of moral problems must always be rooted in a concrete understanding of specific cases and circumstances."[13] As AI is, by its nature, automated, we are faced with the challenge of establishing an ethical framework that works for all cases in general and is computable in an individual case once the AI model is deployed.

While ethicists agree that ethics is about what is right and wrong, there is little consensus on how to assess right and wrong in practical terms, even among professional philosophers of the Western tradition, as demonstrated by the above discussion. Attempts to reduce ethics to the majority opinion are also not satisfactory.[14] The theories summarized above are generally thought to be contradictory. Other traditions than the Western one have yet more theories available, each with its own advantages and drawbacks. Professional businesspeople around the world are challenged to formulate practical ethics guidelines.

Next, we look at the development process of AI and discuss the ethics that might play a role in each step.

The AI Development Workflow

"It's incredibly easy to sell something when you're meeting a real customer need."
"It's incredibly difficult (and often demoralizing) to be asked to persuade someone to buy something that doesn't solve a real problem or fulfill a need."

JOEL PETERSON, *ENTREPRENEURIAL LEADERSHIP*

Technology is more complex than just models and so we must consider the full workflow of technological problem solving in the case of AI:

1 The use case is often selected first. Here we choose what the problem is and what a solution might look like.

2 This must be translated into a technological goal that needs a mathematical formulation of what is good.

3 The model must be trained based on empirical data, which must be collected and cleaned, and the training process is guided by (potentially several simultaneous) goals.

4 When in use, the model is usually the basis for a decision-making process that must recommend the best action based on some criteria.

5 Models are not static but must evolve and therefore must be observed during use, which is known as MLOps or DataOps.

It's instructive here to divide this workflow into two parts. The first part, consisting of the first three steps, is the process of making or training the AI model. The second part, consisting of the last two steps, is the process of deploying the model for applied use, also known as inference. The considerations that come into these parts and their steps are quite different. A technological differentiation is that AI training is generally stochastic, i.e., if you train the model a second time, you will get a similar model but not the identical model. Once the model is made however, the process is fully deterministic, even if the output appears—to the human observer—unpredictable, and the model thus looks like a black box.

It is often said that virtue ethics (Aristotle's theory) cannot provide action guidance. This author begs to differ. The golden mean between two extremes on an axis is a beautiful metaphor for optimization if the concept of this axis is quantifiable. This is why, during the training process, most technology organizations follow the Aristotelian model, at least approximately, because that is the model preferred by mathematics. We define the goodness of a model as being close to the correct answer, not much greater and not much lesser than the right number. As such, we seek the golden mean and call it the "least squares" method. The exact metric, however, is open to debate and this causes some disagreements among AI professionals. One may hear of concepts such as the F1 score, accuracy, mean-square-error, and so on, all of which are ways to measure the position of the golden mean.

During the inference process, however, most technology organizations prefer deontological thinking (rule-based systems) or utilitarianism. The reason is that the AI model output must be transformed into decisions by some post-processing method. In contrast to human decision-making,[15] AI-based decision-making is generally implemented in software where there is no room for interpreting terms or guidelines.

We will now look deeper into each of these five stages.

Use Case

Technology or AI has an explicit purpose. It is purpose-built. That purpose is to give the correct output in as many cases as possible—cases that fit a certain type, of course. A critical element of the purpose is the precise definition of the type of case that the technology is intended to serve.

Correct output can be checked in every single case—we make a measurement in reality and compare it to the model output. However, checking correctness represents a cost and this goes against the usual automation desire of technology. After all, the whole point of AI, in most cases, is to make the real-world measurement unnecessary. So practically, correctness can only be checked statistically! This is a very important difference.

One of the reasons this difference is important is that model errors are also costly, but in different ways. For example:

- False-positives (FP) are when a healthy patient is diagnosed with cancer. The model output is positive for cancer, but this is wrong because the person is actually healthy. In numerical cases, this is an overestimate of the value.

- False-negatives (FN) are when a cancer patient is diagnosed as healthy. The model output is negative for cancer, but this is wrong because the person actually has cancer. In numerical cases, this is an underestimate of the value.

If a model sometimes makes FP and sometimes FN, these errors do not cancel out. Both are mistakes. Both affect people, but these effects differ depending on the case. The FP patient will have to submit to more unnecessary testing and emotional turmoil only to figure out that everything is alright. The FN cancer patient will go home happy but suffer later in time because treatment is delayed.[16] Without performing the above-mentioned check, where a real measurement is taken and compared to the model output, we do not know whether a specific case, to which the AI was applied, is an FP, an FN, or an accurate assessment. We know this only statistically. That fact in itself may be an ethical problem depending on the use case.

These costs are particularly difficult to gauge in low-probability high-cost events, e.g., predictive maintenance for a gas turbine that fails once in 25 years but kills several people when it does so.

Another type of error is applying the model for a different purpose. A face-recognition model trained on Caucasian faces could be very accurate on that type of face. Attempting to apply this model to faces from other ethnic groups is a misuse of the model as it was not built for this purpose. The distribution of FP and FN for such input data is very different than with input data from the original population. This throws off the statistics made for the original population, leading to a performance that cannot be assessed without gathering further ground truth data about the new population.

Finally, we note that some cases are interpreted very differently by different audiences. For instance, the question of disconnecting a life-support system from a critically ill patient was examined. The patient had previously given permission to be disconnected. To the doctors, this was a clear case. To the audience of a review meeting, it was an example of a failure to provide proper care and information to the patient.[17] This example serves to illustrate a central principle in all AI: It is essential to carry a strong discourse between domain experts and data scientists throughout the process to prevent practical failures.

In short, we must examine the use case for ethics. It seems natural to determine its ethical status based on the intentions that we have for it, i.e., to use Abelard's theory, while considering the various potential errors. This is the practical purpose of corporate AI ethics statements: to form a basis on which intentions may be judged. If used correctly, such statements have practical force.

Technological Goal

You always want a good model but generally measure model quality in two different ways: Bias and variance. In throwing darts at a dartboard, **variance** is how tightly the darts are clustered and **bias** is how far off-center the average dart is (see Figure 12.1). In contrast to darts however, AI is generally subject to a trade-off: After a certain point of model optimization, lowering bias increases variance, and vice versa. This means that the designer must choose how much of one versus the other the application can live with. In a way, this is the uncertainty principle of quantum mechanics in AI.[18]

It is important to mention that the word "bias" is used in two different contexts here. The mathematical bias of the previous paragraph means how far from the true value the AI model output is. This bias is a numerical quantity without any value judgment. The second meaning of bias is an attitude of prejudice against some person, group, or things, such as discrimination against women or people of color. It will be clear from the context which version of bias is meant, but it is important to keep in mind that both meanings coexist in AI.

Additionally, we generally want to keep model complexity low, but we also do not want the model to be so simple that it cannot capture the complexity of the data. This is another trade-off. Thus, we find ourselves not only having to make a difficult trade-off but also negotiating between multiple trade-offs. These decisions are ones for human beings and can only be delegated to computers after a high-level decision is made. That is the crucial point at which ethics and responsibility enter the AI training process. What makes both matters even more complex is that each element can be measured in multiple ways and choosing the "right" metric is an art in itself.[19]

FIGURE 12.1 Bias versus variance trade-off illustrated as a dartboard

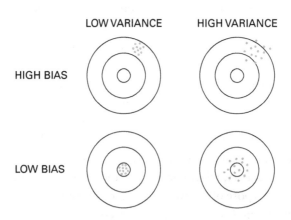

Having chosen the metrics, the search for the optimal point (golden mean) in the multidimensional landscape can begin. Such a search is often not entirely free as certain regions of the landscape are forbidden. These so-called boundary conditions on the search must be enumerated in addition to the metrics before starting the optimization. Simple examples include absolute maximums for bias and variance, demanding maximums for the false-positive or false-negative rate, or requiring the largest residual to be below some cut-off. The last example is particularly tricky for AI but occurs frequently in practical use cases as user organizations desire guarantees on how far off from the target the model can be in the worst case.

As before, the (human) choice of performance metrics and boundary conditions should be based on our intentions for the final application as guided by the corporate AI ethics statement. That statement must be concrete and actionable enough to spell out what is important to the organization so that it can be put into a suitably quantifiable form for a particular use case or for application by the committee of humans in charge of reviewing use cases.

Dataset Generation and Training

Now that we know what we want to do, we start to do it. The first step is to acquire data, which requires deciding what data to acquire, where to acquire it from, and how to do so practically and economically. The first decision is what data, ideally, we want to have. Most datasets come in columnar form. Even computer vision cases where images are analyzed are columnar in nature, where the image itself is one column, with the other columns often called meta-data. Regardless, meta-data is data. Which columns to collect is, ethically speaking, absolutely foundational.

Hidden variables are any variables that are not explicit in the dataset used to train the model. These variables may either not be in the dataset at all or they may be withheld from the training process. However, they are present implicitly in the fact that there are correlations between them and the data. These correlations are generally not causal in nature.

An example is the historical preference of banks to provide loans to men versus women. While the data column of sex may not be present in the dataset, this variable is implicit in many other columns, such as income, or the combination of age and income. The model may then learn the presence of a hidden variable and thereafter discriminate against women, in our example, even though it was never told the sex of any of the people in the training dataset.

Therefore, the bias against women is not explicit since this data is missing and so the discrimination cannot be observed in the normal technological channels. It can only be uncovered later in more complex ways. This is a problem. The solution is to explicitly define the variables that are important to us and to collect the relevant data. This data can then be withheld from training but used in model assessment to determine whether the model is biased. This data could alternatively be explicitly used in model training to enforce equitable treatment of women and men by design.

The challenge is to define precisely which variables we will look at. Sex, race, religion, or sexual orientation are easy data columns to define as ethically important and are, for the most part, also easy to collect in practice. We know about them because they are tightly connected to discrimination against groups that are present in societal debate. In more complex use cases, such as medicine, discrimination against people with a certain gene or preexisting condition are much more difficult in practical reality. Such situations are difficult to gather data about and are also difficult to explicitly define as these groups are not as apparent.

Having collected the right data, the dataset must be cleaned. Measurement mistakes, wrong data, outliers, and such are removed. Missing data may be filled in. These are some standard steps in data science. Ethically, we must examine how biased the dataset is. The dataset may be biased because we collected it in a biased way (e.g., we only asked one kind of person) or because it encapsulates biased human behavior (e.g., bank loans are preferentially given to men). In either case, AI will learn the biases in the dataset, whether the biases are explicit or implicit, and whether they are wanted or unwanted. In the short term, we must additionally "clean" datasets in the sense of removing that bias, but we can only do so practically if the relevant variables are explicitly included and we demand, by design, equity in those variables. In the long term, we would wish for the world to enact structural changes such that future datasets are equitable and do not require such cleaning.[20]

In proud possession of a representative, significant, clean, and unbiased dataset, we start the AI model training process. This is an optimization procedure that attempts to balance between the various extremes mentioned in the previous section. We may now add other trade-offs or boundary conditions to this optimization problem, such as the explicit demand of equitable treatment of various kinds of people or situations—in AI terminology this is called adding terms to the loss function. This is possible because, and only because, we have made the relevant variables explicit. We note that this is inherently an Aristotelian way of doing things. While there are many numerical metrics of fairness, they are incompatible in that we cannot be fair according to all of them at the same time. Therefore, we must a priori choose which metric to obey and which ones to forego.[21] That choice is a human design decision to be made on the basis of intent and founded on the corporate AI ethics statement.

We note in passing that data privacy and the trustworthy stewardship of data is another essential aspect of the journey. While not directly relevant to AI, this is an important issue that poses an ethical risk for the organization and must therefore be kept in mind in any risk-mitigation strategy. As seen in several well-popularized cases, this risk goes well beyond regulatory or legal risks.[22]

Decision Support

Most AI models are used for decision-making purposes. The AI model usually does not supply the final decision as its output but as some numerical quantity that becomes an aspect in the decision-making process. This quantity may be a probability, a numerical value, or a categorization. Using this output to generate a decision requires translating the problem into the input data, retrieving the output data and generating a decision, and deploying the model such that it becomes practical. The decision-making may be fully automated, but this involves pre-processing and post-processing and thus is a larger system than just the AI model.[23]

The various options available for the decision must be clearly identified in advance alongside multiple advantages and disadvantages that can be quantified. Furthermore, each of these measures of advantage must be computable so that we can practically determine them in any case. These can be combined into a measure of goodness for each option and so we may arrive at a ranked list with a clear objective numerical assessment. The best option may then be chosen.

This procedure is often not practical, for two reasons:

1 Human teams often find it difficult to quantify any and all advantage measures of all options, to make them computable, and to make them comparable on a single scale so that the ranking becomes linear rather than a matrix.

2 Many situations have exceptions to any set of such quantifications, which must be dealt with on their own merits. Thus, many decision-making processes involve humans who can selectively weigh the options in specific cases and arrive at compromises.

An element that is often ignored by data scientists is the dynamics of the use case. Primarily, "dynamics" includes two aspects. First, we have the aspect of time, where it may take a long time before it becomes apparent whether the right decision was taken, e.g., it may take years before we know whether a bank loan decision has resulted in a repaid or a defaulted loan. Second, we have the aspect of repetition, in which case the same decision is made multiple times over a long period of time, e.g., medical treatments, in which case short-term upsides in cost might result in long-term downsides of outcomes. The measure of goodness and fairness must be carefully tuned to the holistic desired outcome. An intricate example is admittance decision-making. Discrimination at school admittance influences university admittance, which influences admittance to the job market, which influences career prospects. The chain starts early and once the odds are against any one group, it is hard to recover later.

The choices of fairness metrics, use case timescale, and iterative use are difficult human decisions comparable to our discussion of hidden variables, which must also be identified by a human team well in advance of deployment. There can be no standard default answer, but this requires substantial foresight.

The process of converting the AI model output to a decision or a decision support is the codification of what is usually a human process into software. Human processes have large margins for error because we can consider specific features of the case and deliberate. Software cannot do so. Soft facts or soft rules must be made hard and numeric or must explicitly be reserved for human-in-the-loop decision-making. This is a choice also. These choices are probably best taken based on utility. Measures such as the damage that can be caused in worst-case scenarios and the probability of such scenarios occurring can help in determining utility measures that can calculate where it makes sense to automate decision-making, where a human is best placed, and where exactly the red line in the sand between computer and human must be drawn.

Model Life Cycle Management

The accuracy of a model changes over time. This seems strange at first glance as a mathematical formula does not age or corrode. That model was, however, trained on a dataset that encodes a view of reality, which does age over time. As the world changes, the characteristics of the live data slowly drift from the characteristics of the training data. There comes a time when the model must be adjusted by including current data in the enlarged or modified training dataset. Depending on the use case, this may be once a year or once a month. Data drift may also occur differently, and at different speeds, in different geographies or markets, which may require the splitting of one model into several new-generation models, one for each of these markets. To handle this, models must be continuously observed. It is the AI equivalent of requiring our human physicians to recertify every few years in their chosen specialty.

Observation not only entails gathering the input data that flowed into the model, the output data that the model provided, the decisions that were recommended, the decisions that were actually taken by the human process outside the software system, but also—most importantly—the real-world outcome that resulted from all of this. Clearly, observation is complex and requires preparation to be put in place. Once we have it, we can track our progress. Recall that the fundamental reason for being of AI models is their capability for easy and cheap automation and scaling. The accuracy performance and ethics performance may now be tracked at scale as well.

At this time, we recall the categorical imperative that says that the maxim of what we do should be such that we can reasonably make it universal law. By the very nature of AI and its reason for creation, AI has, now that it is deployed at scale, essentially become a universal law. Moreover, that law is largely automated without being able to make exceptions, unless those exceptions are literally codified, i.e., part of the law.

Laws require oversight and if the oversight determines that the law is no longer in the best interest of the population at large, that law must be changed and perhaps even split from federal laws into state laws. That is the purpose of model life cycle management and the ethics principle of the categorical imperative seems to be its most natural ally.

AI Technology Ethics Framework

"The evidence on whether even teaching ethical principles actually increases ethical behavior is decidedly mixed."

JEFFREY PFEFFER, 7 *RULES OF POWER*

Having discussed some of the challenges in talking about AI ethics and some philosophical bases, this can be put together into a coherent framework, as shown in Figure 12.2.

FIGURE 12.2 The Ethical AI Framework

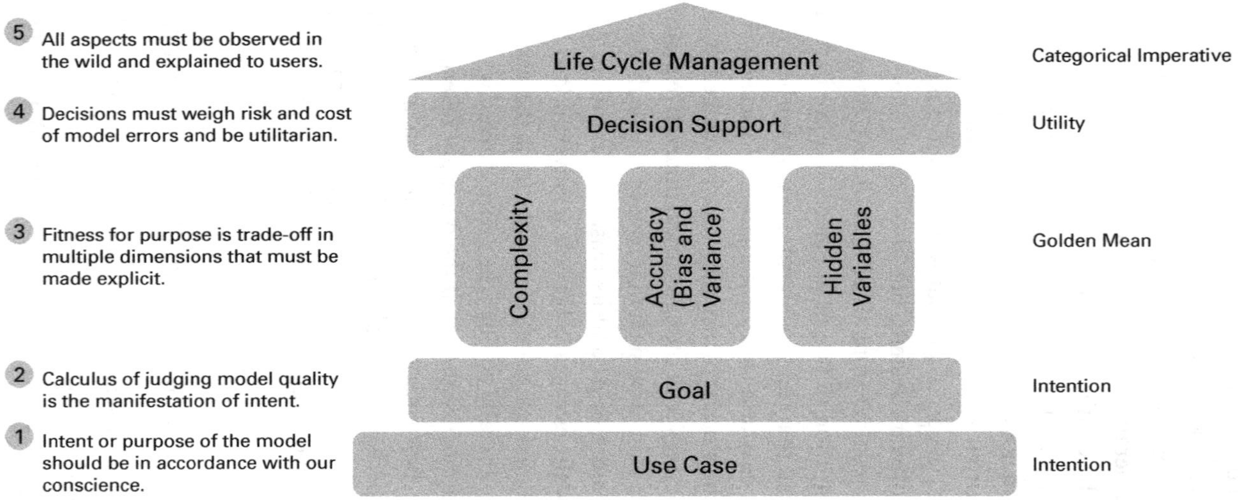

Life Cycle Management

Decision Support — Categorical Imperative · Utility

- Complexity
- Accuracy (Bias and Variance)
- Hidden Variables — Golden Mean

Goal — Intention

Use Case — Intention

5. All aspects must be observed in the wild and explained to users.

4. Decisions must weigh risk and cost of model errors and be utilitarian.

3. Fitness for purpose is trade-off in multiple dimensions that must be made explicit.

2. Calculus of judging model quality is the manifestation of intent.

1. Intent or purpose of the model should be in accordance with our conscience.

1   For the use case, the intention is paramount, and it should be in accordance with our conscience as defined by a corporate value statement. It must also, of course, adhere to the law of the land. In practice, this requires an explicit statement and agreement on what is important to the organization, which in turn requires the formation of a committee that enforces it. The resulting ethical standard must be formulated to be practical in deciding on use cases.

2   In translating this use case to a precise goal, it is again the intention that should be checked as the calculus of how the use case desiderata are manifest in mathematical key performance indicators (KPIs) is developed. The KPIs must be considered carefully in terms of what they will measure and what they will average out. The practical consequences of false-positives and false-negatives must be taken into account.

3   The various trade-offs must be made explicit, and we seek the golden mean. The main challenge is a variety of hidden variables that must be brought from the shadows into the light. Biases can be measured and avoided only if they are explicitly defined, considered, and brought into the dataset. The organization must therefore define all the hidden variables that it will explicitly look out for, gather data about, and require equitable treatment for.

4   Once the model is ready for decision support, it is the utility that rules its ethics. The greatest good for the greatest number seems a good goal for decision-making. However, this presupposes an explicit calculus must be defined and documented.

5   Finally, the life cycle management that observes the model in practice and provides explainability to the users should obey the categorical imperative. The maxim of what you do—the recommended decision—should be good enough to be used universally by everyone who fits the relevant profile of the example at hand.

Practical Action Items

"People will do something—including changing their behavior—only if it can be demonstrated that doing so is in their own best interests as defined by their own values."

MARSHALL GOLDSMITH, *WHAT GOT YOU HERE WON'T GET YOU THERE*

Theoretical discussions are nice, but we now distill it into some practical recommendations that organizations might actually implement to get this done. Why should they do it? Apart from being good global citizens, the practical, economic, and business reason is risk management. Ethical failures, bias, and discrimination can be quite costly and publicly embarrassing for people and organizations, not to mention occasionally illegal. Avoiding this can be seen as an insurance policy, which comes at a price.

AI teams should lean into being an ethics review board that draws up the required ethical baseline documents and goes through these stages with the engineering teams. Many projects are not as linear in their development as described here, but all have these elements in the agile software development process that they enact. Examine each element on its own merits so that all of it makes sense and can be defended.

Make your hidden variables explicit and document everything you decide and the reasons why you decide it. Transparency does not guarantee ethics, but ethics demands transparency. As we mathematicians would say: Transparency is a necessary but not a sufficient condition for ethics.

Finally, it is necessary to reflect on all aspects of the project in detail and to document these reflections for the whole team, future members of the team, management, users of the technology, and the general public.

Skipping these steps is a technical debt that has a high interest rate! Briefly, these are the next steps for any organization that wants to take this seriously:

1 Form an AI ethics board. This should have data scientists, domain experts, lawyers, businesspeople, and professional ethicists (sometimes called embedded ethics)[24] on it. It should also be sponsored by a C-level executive and outfitted with authority to make changes to projects and the requirement that all relevant projects be reviewed.

2 Write an AI ethics statement. This statement should not contain general platitudes but enact specific measurable rules and desiderata based on which projects can be assessed. Terms must be defined. Goals must be quantifiable. Standards must be spelled out in detail. An excellent manual on how to do this is *Responsible AI* by Olivia Gambelin.[25]

3 Assess the use case behind every potential application based on the AI ethics statement. Is the intent in accordance with corporate values?

4 Assess the AI training loss function, trade-offs, and boundary conditions as well as the AI model goodness assessment criteria by their intent based on the corporate ethics statement. Are we solving for the right thing?

5 Decide on the various dimensions that the project will demand equity on and define precisely what kind of data must be collected to represent them. Decide whether this equity will be explicitly required by including it in the AI training loss function, or whether the equity will merely be assessed after training is complete. Implement data governance that collects, secures, and maintains the collected data and meta-data, including the relevant ethical bias variables. Then train seeking the golden mean of all the goals.

6 Design the post-processing of the model for decision support in such a way that the decision results in the greatest good for the greatest number. In this design process, spell out precisely and numerically how the good and the number affected are going to be practically measured or calculated. Put into place rules that may or may not be required to make sure that edge cases are properly included in this utility scheme.

7 Perform model life cycle management and model observability so that the decision can reasonably hold as a universal law for all cases of the type under consideration in this use case. Think of exceptions to which this may not apply.

8 Document all these decisions, changes, and reasons for them.

All these actions are to be performed by the ethics board in collaboration with the teams behind the use case, software development, or AI model development. This may seem overly burdensome but there are several significant benefits: (a) the application will be much more likely to serve its true purpose after all this review, (b) the model will be more accurate and address the right issues, (c) the model will not discriminate against any groups that we do not wish to discriminate against, and (d) the risk of public embarrassment or legal action is significantly reduced.

KEY TAKEAWAYS

1 While "AI ethics" and "responsible AI" remain flexible terms, it is helpful to define what these mean to your enterprise, including concrete steps taken to ensure that these standards are being upheld.

2 Create a committee whose job it is to watch and enforce these standards on both purchased software and internally developed models and tools.

3 Make hidden variables explicit if they help to quantify and prevent unwanted bias, which presupposes clarity on which dimensions you want to be equitable.

Notes

1 Both of these examples are discussed here: https://syncedreview.com/2021/01/01/2020-in-review-10-ai-failures/ (archived at https://perma.cc/K2SE-Z4U9)

2 Calma, J. (2022). AI suggested 40,000 new possible chemical weapons in just six hours. The Verge. www.theverge.com/2022/3/17/22983197/ai-new-possible-chemical-weapons-generative-models-vx (archived at https://perma.cc/68NT-6VXH)

3 Yucer, S., Tektas, F., Al Moubayed, N., and Breckon, T. (2024). Racial bias within face recognition: A survey. ACM Computing Surveys. https://doi.org/10.1145/3705295 (archived at https://perma.cc/R4NT-JSBK)

4 Sidgwick, H. (1962) *The Methods of Ethics*, Macmillan.

5 Brynjolfsson, E. (2022). *The Turing Trap: The promise and peril of human-like artificial intelligence*. Dædalus.
Acemoglu, D. and Restrepo, P. (2021). Tasks, automation, and the rise in US wage inequality, NBER Working Papers 28920, National Bureau of Economic Research, Inc.

6 Yu, K.-P., Tao, J., and Ivanhoe, P.J. (2011). *Taking Confucian Ethics Seriously: Contemporary theories and application*. SUNY Series in Chinese Philosophy and Culture.
Graham, A. (2004). *Later Mohist Logic, Ethics, and Science*. The Chinese University of Hong Kong Press.
Lee, J.H. (2014). *The Ethical Foundations of Early Daoism: Zhuangzi's unique moral vision*. Palgrave Macmillan.
Nadkarni, M. (2011). *Ethics for Our Times: Essays in Gandhian perspective*. Oxford University Press.
Srivastava, C., Dhingra, V., Bhardwaj, A., and Srivastava, A. (2013). Morality and moral development: Traditional Hindu concepts. *Indian J Psychiatry*, 55, S283–S287.
Garfield, J.L. (2021). *Buddhist Ethics*. Oxford University Press.
Harvey, P. (2000). *An Introduction to Buddhist Ethics: Foundations, values and issues*. Cambridge University Press.

7 Gupta, A., and Heath, V. (2020). AI ethics groups are repeating one of society's classic mistakes. *MIT Technology Review*.

8 Aristotle, *Nicomachean Ethics*, 350 BC.

9 Spade, P., and Abelard, P. (1995). *Ethical Writings*. Hackett Publishing Co.

10 Mill, J.S. (1879). *Utilitarianism*. Longman's, Green, and Co. *Stanford Encyclopedia of Philosophy*, https://plato.stanford.edu/entries/utilitarianism-history/ (archived at https://perma.cc/ZP4S-GT9X)

11 Kant, I. (1788). *Critique of Practical Reason*.

12 Childress, J. (2009). Methods in bioethics, in *The Oxford Handbook of Bioethics*, pp. 15–45.

13 Jonsen, A. and Toulmin, S. (1988). *The Abuse of Casuistry*. University of California Press.

14 LaCroix, T. (2022). Moral dilemmas for moral machines. *AI and Ethics*.

15 Childress, J. (2009). Methods in bioethics, in *The Oxford Handbook of Bioethics*, Oxford University Press, pp. 15–45.

16 Kahneman, D., Sibony, O., and Sunstein, C.R. (2021). *Noise: A flaw in human judgment*. Little, Brown and Company.

17 Kaufert, J. and Koch, T. (2003). Disability or end-of-life: Competing narratives in bioethics. *Theoretical Medicine*, 459–469.

18 Banger, P. (2021). *Machine Learning and Data Science in the Oil and Gas Industry: Best practices, tools, and case studies*. Elsevier.

19 Reinke, A. (2022). Metrics matter—how to choose validation metrics that reflect the biomedical needs, in 3rd International Conference on Medical Imaging and Case Reports (MICR).

20 Sargent, S.L. (2021). AI bias in healthcare: Using ImpactPro as a case study for healthcare practitioners' duties to engage in anti-bias measures. *Canadian Journal of Bioethics*, 4(1), 112–116.

21 Dwork, C., Hardt, M., Pitassi, T., Reingold, O., and Zemel, R. (2012) Fairness through awareness, in Proceedings of the 3rd Innovations in Theoretical Computer Science Conference (ITCS).

22 Winlo, C. (2020). The 10 data privacy fails of the decade—and what we learnt from them. techradar.pro, www.techradar.com/news/the-10-data-privacy-fails-of-the-decade-and-what-we-learnt-from-them (archived at https://perma.cc/5DAM-N6Q8)

23 Fazelpour, S., Lipton, Z.C., and Danks, D. (2021). Algorithmic fairness and the situated dynamics of justice. *Canadian Journal of Philosophy*, 1–17.

24 McLennan, S., Fiske, A., Tigard, D., Müller, R., Haddadin, S., and Buyx, A. (2022). Embedded ethics: A proposal for integrating ethics into the development of medical AI. *BMC Medical Ethics*.

25 Gambelin, O. (2024). *Responsible AI*. Kogan Page.

13

Governing and Maintaining AI Models and Applications Over the Long Term

The most important decisions in life are the actions you choose *not* to take. The spirit of **governance** is to establish some ground rules around what the enterprise will refrain from doing, what it will require, how that is decided, and who gets to make those decisions.

To be efficient in deciding what not to do, it is helpful to preempt lots of case-by-case decisions by agreeing on some general principles or values that guide your decisions. If you find that a particular situation is simply an example of such a principle, you do not really need to decide but merely instantiate an already decided outcome.

Setting Your AI Principles or Values

"After forming your beliefs you then defend, justify, and rationalize them with a host of intellectual reasons, cogent arguments, and rational explanations. Beliefs come first, explanations for beliefs follow."

MICHAEL SHERMER, *THE BELIEVING BRAIN*

Most enterprises already have certain values or principles in place that are a guiding light for the whole company, whether it concerns AI or not. Those are great starting points. It is an important and illuminating discussion at the leadership level to decide on a few broad and sweeping desiderata that you want to hold true in all cases. It may remind us of Kant's categorical imperative: Act in such a way that the maxim of your action ought to be a universal law—in this case, universal for your enterprise.

This is a very open-ended discussion, and a great many possible principles or values may make your list. This book cannot discuss them all and so I provide an example set of values from the company I work for in Table 13.1. This set was decided on by a committee of senior leaders from IT, legal, compliance, security, operations, and AI, and approved by the heads of the business units.

TABLE 13.1 An example set of six principles guiding AI governance

SAFETY	This includes designing resilient AI systems that function consistently under various conditions, ensuring reliable outcomes and protecting people, systems, and data from harm, errors, and attacks.
All AI users and developers adhere to high standards, ensure transparent AI systems, and comply with evolving regulations. Defined roles help manage AI development and use.	**ACCOUNTABILITY**
PRIVACY	The enterprise commits to handling all personal and sensitive data responsibly, minimizing personal data processing, and using robust cybersecurity measures while complying with relevant laws and regulations.
AI practices optimize resource use, have clear operational impacts, and are environmentally responsible. AI projects drive business value and have clear performance metrics.	**EFFICIENCY**
HUMAN-CENTRICITY	AI augments human judgment, maintains human oversight, and actively mitigates bias. AI systems empower users and allow human interventions when necessary.
Keep AI applications running 24/7 when required for business operations. Models, applications, and infrastructure will be designed with appropriate disaster recovery in place.	**RELIABILITY**

Having a set of five or six principles is a good number. A little less would work well too, but more probably becomes unwieldy.

While Table 13.1 supplies basic definitions, it will be necessary to provide more detailed definitions of these values so that everyone is clear on what AI applications or methods each principle forbids or desires. In case a principle is violated, it will become important to be able to measure the impact of the consequences, as we will see in the next section on risk management.

These six principles form a high-level decision-making framework for approving a use case or feature of an AI application. The development process requires many decisions to be made, some of them seemingly minor, that may have macroscopic consequences. For example, when the enterprise aims for safety, is this primarily targeted at the forklift not running people over in the factory, providing psychological safety in the office, or preventing identity theft online? If the answer is all of the above, some priority order should be established. After all, these principles should not remain theoretical slogans but be directly convertible into actions.

Conducting AI Risk Management

"To stay ahead, the thinking goes, a company must stake out a distinctive strategic position—doing something different than its rivals. [...] Our research shows that simple managerial competence ... should be treated as a crucial complement to strategy."

RAFFAELLA SADUN, NICHOLAS BLOOM, AND JOHN VAN REENEN,
WHY DO WE UNDERVALUE COMPETENT MANAGEMENT?

Looking across the entire enterprise, there will be a multitude of AI-powered applications licensed from vendors, ordinary applications with AI-enabled features, explicit AI models being used directly, in-house developed AI models and applications, and various service offerings either using or making AI applications. Any of these may be used in more than one way; that is, for more than one use case.

Most of these applications and use cases pose little risk to the business, but a few may pose a higher risk. As AI models are inherently never 100 percent accurate, and many serve as the basis for decision-making that may have real-world consequences, it is important to manage the cumulative risk of the portfolio of AI applications and use cases to protect the enterprise.

An established method of managing risk is the risk matrix. Each risk item is assessed on two dimensions—probability and impact—and so plotted on the matrix. That immediately raises three questions: What is a risk item and how should probability and impact be determined?

In AI, a risk item is when the AI provides an incorrect output, and this drives a suboptimal decision or outcome. If AI provides the correct output and a suboptimal outcome occurs, one might assume that this is due to some other driver, but not AI, and so this will not be included here. Depending on the use case, a single wrong AI output may result in multiple bad outcomes depending on the process that follows the AI output.

To be clear on this, the medical diagnostics example will shed some light. A healthy person is mistakenly diagnosed with cancer by the AI system. What are the possible outcomes of this? There could be many possibilities, but here are some major ones. First, the mistake is caught by a human doctor and so no bad outcome results at all. Second, the patient or doctor orders a second opinion that finds the error and this has caused temporary frustration. Third, somehow the error leads to unnecessary treatment that causes actual harm with consequences. These three situations contain two risk items since the first situation does not lead to a bad outcome.

For each of the risk items, the probability of occurrence must now be determined. This is clearly not just the probability of the AI model providing an incorrect output but must be multiplied by the probability of the downstream process also producing the bad outcome. In the above example case, the probability of the last risk item is the probability of the AI system providing the wrong answer, the human process not catching the error, and the human process ordering and administering treatment.

For each instance of model inference—in this case for each individual diagnosis—the probability is likely quite low. However, your enterprise is likely to use the AI application many times. To obtain the probability of this event happening to the enterprise over a certain time frame, such as a year, the individual event probability is now multiplied by the number of inferences over that time frame. Due to the low unit cost of AI inference, this number may be large. Therefore, the overall probability could be high. For example, if the individual probability is 1 in 1 million, but the model will be used on 100,000 cases per year, the probability overall is 1 in 10, or 10 percent per year!

Risk is an example where the scalability of AI has surprising ramifications.

For a risk matrix, you may choose to divide risk into five major segments by probability, as shown in Table 13.2. As discussed, you must provide the

TABLE 13.2 Possible assignment of risk categories by probability

Rare	1% probability
Unlikely	10% probability
Possible	50% probability
Likely	90% probability
Almost Certain	99% probability

right context for the probability, such as the number of cases you are assuming and the time frame for that estimate. Only then do probabilities make sense.

The second dimension of the risk matrix is the impact of the event. The impact should be assessed by using the principles you established in the previous section and ultimately converting the situation into a financial impact so that the various situations become comparable. It is important to note here that I do not recommend using money out of a cold-blooded profit motive, not caring about emotional or other impacts. Far from it. The point is to establish a single numerical metric by which different, and otherwise incomparable, situations can be compared, albeit perhaps superficially.

Continuing the above example of the six principles provided, what follows are possible impact definitions for those principles. As with probabilities, the impact dimension is broken up into five categories:

- **Insignificant** impacts are those that cause no injuries to personnel or damages to assets (safety), have no impact on reputation (accountability), cause no loss of personal or business information (privacy), cause no loss of business or production capacity (efficiency), and no impact on the wellbeing of enterprise employees or contractors (human centric) valued at below $10 million.

- **Minor** impacts are those that have some impact in these dimensions valued at between $10 million and $100 million.

- **Moderate** impacts are those that have major injuries to personnel or damages to assets, or localized impacts on reputation, information, production, or wellbeing valued at between $100 million and $500 million.

- **Major** impacts are those that cause a single fatality or destruction to a single asset, a major national impact on reputation, major loss of information, of production capacity, or a major impact on wellbeing in a single business unit valued at between $500 million and $2 billion.

- **Catastrophic** impacts are those that cause multiple fatalities or destruction of multiple assets, a major international loss of reputation, extensive or global loss of information, extensive or international loss of production, or extensive impact on wellbeing in multiple business units valued at over $2 billion.

Having assessed a risk item in both probability and impact, the item is now placed on the risk matrix. An example of such a matrix is given in Figure 13.1.

The placement of risk items now triggers a review, by a governance committee, of the risks incurred with two aims. First, the committee designs mitigation measures to lower both probability and impact. Second, the committee decides whether the risk is worth taking. If the risk is deemed not worth taking, the AI application is not approved for deployment, unless perhaps it can be made significantly less risky by doing more training or doing something fundamentally different.

Again, the ultimate aim of governance is to decide not to do things. By default, applications and models, of course, are made to be used and naturally transition step by step towards deployment. The last stage before an application is turned on for live use is the governance review process.

FIGURE 13.1 A risk matrix example

Potential Impacts	Catastrophic	Moderate	Moderate High	High	High	High
	Major	Low	Moderate	Moderate High	High	High
	Moderate	Very Low	Low	Moderate	Moderate High	High
	Minor	Very Low	Very Low	Low	Moderate	Moderate High
	Insignificant	Very Low	Very Low	Very Low	Low	Moderate
		Rare	Unlikely	Possible	Likely	Almost Certain
		Likelihood of Occurrence				

The kind of review should depend on the risk and impact level—that is the purpose of the risk matrix. Items with very low or low risks can be governed by the central AI team's governance function itself. Moderate risks may need to be brought to the attention of a corporate risk function, if such exists in your enterprise. Risks that are high will need approval from a higher-level governance committee, likely including C-level leaders of the enterprise. The committee, therefore, depends on the risk level of each risk item. Distributing the approvals process in this way is a matter of expediency as senior leadership cannot be expected to participate in every decision. In a moderately large enterprise, it must be assumed that 5–10, or more, AI-related risk items must be discussed every week and so this process must be efficient.

Governing AI

"It turns out that when you try to multitask, not only do you waste more time [...] but you end up doing both tasks less well."

RAHAF HARFOUSH, *HUSTLE & FLOAT*

AI governance is frequently acknowledged as highly challenging in enterprise environments because it requires coordination and agreement from all lines of business as well as the pooling of oversight since AI is connected to many parts of the company. As an active area of practice and research, there are many opinions on how to do it. In spirit they all more or less agree, and the differences are really in operational details and points of emphasis.

Having served on the Artificial Intelligence Safety and Security Board (AISSB) of the US Department of Homeland Security, I will here present, in my own words and with my own opinions, the coordinated view of that board as published in 2024.[1] A summary of the key aspects is given in Figure 13.2 where it is clear that governance is treated in five nested layers of desired qualities.

Security and Cybersecurity

At the core level, the physical infrastructure of computers and other equipment in data centers as well as the physical and digital access to this infrastructure must be secure. While physical security is clear, digital security

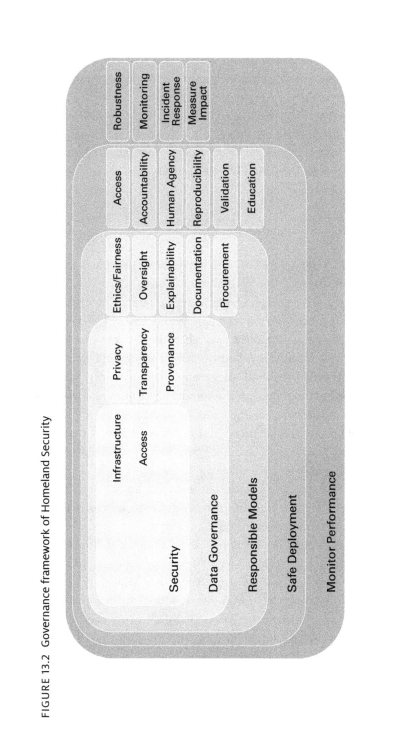

FIGURE 13.2 Governance framework of Homeland Security

is complex and evolving daily. Cybersecurity is a dedicated, specialized, and essential discipline that is considered a prerequisite for doing complex computational work like AI on proprietary data. Having staff in IT dedicated to this is necessary.

Building an AI program does not create the need for cybersecurity—that need has been there all along since the regular software and data systems need protecting. However, due to the unique features of AI of scalability and influence on the entire enterprise, the impact of any security vulnerabilities does increase significantly with AI and so underscores the need to take this seriously. Perhaps one can think of this as increasing your insurance coverage after buying a sports car.

Data Governance and Documentation

The second layer addresses data governance specifically in three areas: privacy, transparency, and provenance.

Data privacy is a complex topic heavily intersected with evolving governmental compliance laws that can be very different in different countries and that can change significantly over short timescales. The basic idea is to protect the rights of individual human beings by collecting, storing, and making available data about them only when necessary. Specifically for AI model training and inference, it is almost never necessary to know the identity of an individual person. Virtually all AI models are models of large groups of people who are grouped together by relatively superficial similarities such as location, age, or buying behavior. As pointed out in Chapter 12, however, the exclusion of certain variables can actually serve to accentuate unwanted bias. So, choosing the elements that are included and excluded in a dataset, relative to a desire for privacy and avoidance of bias, is not an easy matter and generally depends on the use case.

Being transparent as to what data is collected and how it is processed is considered a core principle. Depending on laws, consent may be required for collecting certain pieces of data. There is a slippery slope between the extremes of burying this transparency in the small print of a book-length end-user license agreement that virtually all of us click through without reading and communicating it explicitly. Therefore, there are many levels of transparency that the enterprise may choose to adopt.

Internally, for citizen data scientists and AI experts to realize what data has been collected and what it all means is a crucial prerequisite to any modeling and analysis. Without this knowledge, most efforts at interpretation will be misguided. Careful documentation is therefore in the interests of the enterprise itself.

Provenance is the documentation of where the data came from, how it was collected and transformed. The term comes from the art world where provenance is the collection of documents that prove the authenticity of the artwork from a certain artist and the various purchases until the present moment tracing its ownership and whereabouts. That is very much the spirit of data lineage as well: Tracing the event when the data was collected and then moved and transformed through various IT systems until it arrived in the AI model's training, testing, or inference datasets. This influences the interpretation of the AI model output and therefore the decision-making process resulting from it. Largely, provenance is related to the uncertainty of the model output and therefore the trust that the human decision-maker has in the conclusion.

Responsible Models: Doing the Right Thing

The third ring in Figure 13.2 wants to ensure that the AI models are responsibly made. Five main points cover the creation of models: Ethics and fairness, oversight, explainability, documentation, and procurement.

Ethics and fairness was covered in more depth in Chapter 12. Briefly, models must not discriminate against any groups of people and must not pursue agendas that are harmful to people. What that means in detail and how to ensure that is very complex indeed and, as noted, is the topic of many books in its own right.

Humans having oversight over the models involves having a robust review and approvals process in place to make sure that AI models and applications are checked and intentionally released into the enterprise or to its partners and customers to accomplish some beneficial purpose. In practical terms, the enterprise must form a committee having this oversight and being accountable for the judgment calls made.

The ability to explain the AI model outputs may or may not be necessary for specific use cases. However, most of the time, users benefit from at least being given a precise and relevant indication on how accurate the output is and some of the context in which this output must be interpreted to yield a decision. Strictly speaking, explainability refers to a technical matter in which the transition from the model's input to the output is explained in such a way that the human reader can understand it. It is often unclear how to reach such understanding in a world where many models have an extremely large number of parameters and effectively constitute a black box.

In this context, the requirement of documentation refers to the need to document how the model was made. This primarily involves the steps taken to curate, clean, and transform the available data into the training dataset. The decisions on the model type and its tuning and training are involved also. Lastly, the ways in which the model was tested and ultimately deemed good enough for practical use are included. This documentation serves at once as a justification for why the model has been released into the wild and a tutorial for anyone wanting to maintain or build upon that model. It also serves to instill trust and confidence for any user.

Responsible procurement of AI tools and services is important not just from a corporate governance and compliance point of view but also from a narrow AI point of view. The landscape of available tools and services in AI is quickly evolving, large, and quite obscure. Deciding what to buy and then documenting the steps taken during that decision is good practice in navigating this landscape. Investigating which offerings are genuine AI with good technical methods and which others are using AI as a buzzword, using poor methods, or promising outcomes that are not realistic and so on, can be time-consuming and may require high technical competency.

Safe Deployment: Doing the Thing Right

The fourth ring now looks at deploying the AI models well and discusses this in six topics: Access, accountability, human agency, reproducibility, validation, and education.

A core element of governance is to provide access to the AI application only to those who need it. Often that has less to do with the AI model and more to do with access to the data being processed by the model. AI applications are deployed to accomplish specific tasks. The people who need to do those tasks should have access and not others. Identifying who that is in your enterprise for each use case and having a process in place to add or remove people from that access list is an important governance process. This can be managed easily by having in place explicit user access lists for each use case managed by the deployment software of your AI models.

In case something goes wrong, having identified who has accountability for it is another core aspect of governance. Most AI models can just be turned off when trouble occurs, but this is not always the case, depending on how deeply integrated the model might be. Even if the model is temporarily turned off, someone must fix whatever went wrong and discover the root cause. This may be as simple as having a named owner for each model and application who has the responsibility of knowing all about this, so that when the time comes they are able to act quickly.

Having human agency implies the desire that there be a human in between the AI model and an action in the real world. This line is blurred by some AI models, and particularly the drive towards agentic AI recently. The danger at hand is that a wrong AI model output is thus transformed—automatically and instantaneously—into real damage. To what extent the enterprise will allow this and what non-AI guardrails need to be put in place to prevent grievous harm is covered in this element, which can lead to some deep and lengthy debates.

Reproducibility is a scientific requirement that the same input should yield the same output. For numerical models, that is usually built into the model itself. For large language models, that is famously not the case, but mostly the outputs are quite similar, at least in content. The other side of reproducibility is that models evolve over time as they are retrained through multiple versions. If the same input is provided to two versions, generally the answer will not be identical but one would want it to be quite close. Maintaining such behavior over an evolution of models is difficult and will generally not be true.

Validation is another scientific desire that the model's outputs have been checked and found to be good in most cases. Generally, the few cases where the AI model provided poor outputs are investigated in more detail to determine the extent to which the model can be trusted. It is important—as noted several times in this book—that we know the distribution of inputs for which the model is good. Validation is the governance step at which this is formalized and documented.

Education rounds this out through providing this information to the stakeholders and users in a way that they can understand and keep the limitations in mind while they use the application and make decisions and take real-world actions on its basis. The level of education and training must be appropriate to the use case and change depending on what is expected from the users and the risk level of the application.

Monitor Performance: Keeping a Weather Eye

Once the application is turned on and the model is in live use, the fifth and last ring of governance is activated. This has four elements: robustness, monitoring, incident response, and measuring impact.

Related to reproducibility is the desire for robustness, which is the expectation that when the inputs are slightly different, then the outputs should also only be slightly different from each other and not qualitatively different. This is *not* guaranteed! In the sense of the mathematical field of chaos theory, that fact generally makes AI models chaotic. As chaos theory is quick to point out, the boundaries between areas where small differences in inputs lead to small differences in outputs are complex and so cannot be practically tracked. In some applications, additional rules must be put in place to prevent too-large changes being made at any one time. This again highlights the fact that AI models cannot be used by themselves, but they must reside in several layers of non-AI deterministic software that filters out many of the sticky inputs and outputs.

Monitoring models will be discussed in more detail in the next section on maintaining models. It is necessary because the distribution of input data generally changes over time and so the accuracy of the model decreases over time. Gaining visibility on this and then retraining the model to recover accuracy is an essential part of the workflow.

If something does go wrong, the enterprise must respond to the incident. This could be as simple as turning the AI off for the time being or as extreme as dealing with the press and the public. Depending on the risks involved, the governance team and the AI team must have some basic responses planned out so that they can respond quickly and decisively to limit the damage done.

Finally, the impact of the AI ought to be measured. The very reason for doing AI in the first place is, of course, the hope that the impact will be large, ubiquitous, and positive. The enterprise will want to see explicit evidence that this is the case, and those tracking mechanisms should be put in place from the start with numerical and well-defined metrics. The isolated cases where the impact may be smaller or, hopefully rarely, negative should also be detected by these mechanisms so that the AI team can work on mitigation strategies. Many companies "feel" that AI is beneficial but do not track it and so do not know for sure or how much. Tracking AI impact is not easy because there are secondary effects in multiple places that may detract from the benefits but are harder to measure or tie into a holistic view.

ML Operations: Maintaining Models

"Here's my theory: Disengagement is the issue underlying the majority of problems I see in families, schools, communities, and organizations and it

takes many forms ... You disengage to protect yourselves from
vulnerability, shame, and feeling lost and without purpose."

BRENÉ BROWN, *DARING GREATLY*

Like any machine, AI models also need maintenance. Having been trained, AI models can be used and do not change due to being used. They do not wear out or become worse in some way but remain static. As pointed out multiple times however, the world does change and so the distribution of input data changes. Thus, there are two intersecting loops of AI model creation and maintenance displayed in Figure 13.3 that will be discussed in this section.

The top loop starts at arrow 1 and returns there through arrow 7. The empirical data is generated by some data acquisition at edge devices and this data is transferred into a database (arrow 1) into which a human domain expert provides some annotations or labels (arrow 2). It is then fed into an AI training algorithm (arrow 3), resulting in an AI model (arrow 4) that is then assessed (arrow 5). The human AI expert assessing the model reaches a decision on whether the model is good enough or not (arrow 6) and sends it back to the beginning if it is not good enough (arrow 7). To make the model better, there are generally three main options: get more data, train with more resources (such as hyper-parameter tuning or more training epochs), or do more science by changing the model architecture.

If the model is deemed good enough, it enters the second loop by being packaged in software (arrow 8) that makes the model executable and so it is served to the live data acquisition system at the edge (arrow 9). After this, the outputs are monitored (arrow 10). The monitoring is usually automatic, but occasionally new domain expert inputs in the form of annotations and labels are needed (arrow 11). This information leads to a model assessment (arrow 12) as before. If the assessment is negative, the model goes back to the model-making loop as a new version of the model is now needed.

If the assessment is positive, the model may continue through the inference stage (arrow 13) at which an explainability engine may be used (arrow 14) if needed, which is then provided back to the device that runs all the software (arrow 15). Finally, all this information can be provided to the end user as output (arrow 16).

The decision point (arrow 6) is the critical step that decides to either ship the model or train it again. This is where governance happens and determines whether the model is good to go. It is advisable to institutionalize a formal approval process at this point, as discussed above.

FIGURE 13.3 The machine learning operations (MLOps) life cycle

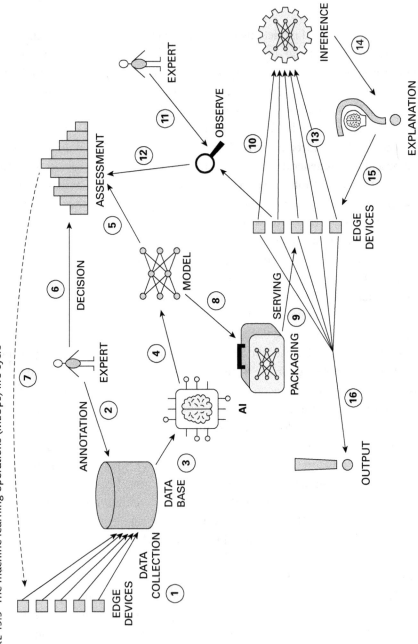

These interlocking loops are known as MLOps, or machine learning operations. These loops keep going in perpetuity or until the application is no longer required. As noted before, in the retail industry the loops are fully traversed once per week as consumer behavior changes on that timescale. Most use cases are slower than this, but it is important to realize that the speed of MLOps is determined by the world. The AI team or the enterprise has no influence over this.

It is important for the enterprise to fully understand the necessity of MLOps because it represents a cost in both effort and money that will remain stable over the entire lifetime of the use case.

AI Application Management

"Programs must be written for people to read, and only incidentally for machines to execute."

HAROLD ABELSON, *STRUCTURE AND INTERPRETATION*
OF COMPUTER PROGRAMS

The AI model is wrapped in multiple layers of software making up the full application. While the AI model's accuracy will degrade and require regular maintenance because of the world's evolution, the application also needs maintenance. While this maintenance is not related to AI, it is nonetheless important to derive value from AI because it is directly related to user adoption.

Users do not use the AI model. They use the application that contains the AI model. In my experience, nearly all the comments that users provide about AI applications are related to user experience, the user interface, and the manner in which the data is collected and presented. These comments get made whether the AI team is there to hear them or not—answering that Zen question about whether users make a noise near the water cooler when no developers are around to hear them: yes, they do.

The AI team should have a way to collect regular feedback from the user groups on each use case and to communicate back to that user group the plan for evolving the entire AI application—including the model but also including the interface and usability. They must know that their feedback is heard, taken seriously, and ultimately implemented to retain users. Remember that there are no unhappy users.

These regular touchpoints can take on different forms depending on the use cases. For use cases that are internal to the enterprise, you are likely to have a definitive list of users available who are all fellow employees of the same enterprise. Therefore, you are, at least in theory, capable of speaking to them. In fact, most enterprise AI applications have a small number of users so that conversations are practical. For example, specialized chatbots for legal contracts or financial statement analysis may have a few tens of users or less. Predictive maintenance forecasts are consumed by maintenance planners who probably make up a small group, and so on.

On the other end of the spectrum, if your user group is very large—such as consumers for a retail website—you can embed statistics collection into the website and track their usage numerically. Feedback can be collected by A/B testing, which means that you roll out a change only to some users, the A group, while the B group keeps the current version. Based on the statistics, you can then objectively decide whether the change was an improvement. Many popular retail websites famously perform such test runs on a daily basis.

Application maintenance needs attention and staff. It represents another lifetime cost for a use case and must not be ignored if the use case is going to drive business value in the long term.

KEY TAKEAWAYS

1 Carefully choose the AI principles or values that your enterprise cares about and perform use case and application risk management on the basis of those principles.

2 Introduce comprehensive governance, for example along the lines of the Homeland Security recommendations discussed here, with the key objective being to have a solid process for deciding what not to do.

3 Keep in mind that both AI models and applications must be maintained in perpetuity, which is a regular cost to the business requiring dedicated staff. With every finished project, the portfolio of models and applications being maintained grows.

Note

1 Homeland Security (2024). Groundbreaking framework for the safe and secure deployment of AI, press release, November 14, www.dhs.gov/archive/news/ 2024/11/14/groundbreaking-framework-safe-and-secure-deployment-ai-critical-infrastructure (archived at https://perma.cc/5SHQ-QZ9S)

14

Managing Vendors and Encouraging Open Innovation

Buy Versus Build

"An investment in knowledge pays the best interest."

BENJAMIN FRANKLIN

You cannot do everything by yourself. The field of AI is full of offerings of all kinds, from market and academic research, via software and hardware products, to service offerings for every possible skill you could want to have. The companies supplying all of this are many and changing daily due to a vibrant ecosystem of startups funded by venture capital and private equity and eventually bought out by strategic investors.

Before delving in and outsourcing development, it is expedient to decide on a buy-versus-build strategy for the whole AI program at your enterprise. This will cover what sort of elements you will buy and what others you will build in-house with your own staff. It will cover the standards you will apply to different vendors and how they will be vetted and chosen.

In the case of software products, buying is generally cheaper and faster than building because those products are ready and past users have already worked out the bugs with the vendor. The features are unlikely to fit your needs perfectly and so either you will have to make do or contract with the vendor to offer some customization.

In the case of services, you are buying access to ready-made knowledge, established skills, fast access to talent, or simply outsourcing work. If the service you need is transactional, has a definite end date, or needs to be started soon, delegating it to a vendor is quicker, cheaper, and hassle-free compared to hiring new staff.

I recommend buying what you can and building only what you must. Why must you build certain things? You may not want to release sensitive data. Strategic intellectual property should be retained internally. Differentiating products ought to remain in-house.

From the entire complex life cycle of an AI application (discussed in Chapter 7) and the various tools needed to create and maintain this application, it is the data, the model, and the workflow that are the central pieces. Depending on how differentiating and strategic the use case is, you want to retain this expertise on your staff. The tools necessary to curate the data, to train and monitor the model, and to interoperate with the workflow can all be bought without significant liability to the enterprise. The work to build all the integrations can be outsourced.

Even AI models themselves can often be obtained externally. Large language models are an obvious case in point. It is highly unlikely that you will want to train your own LLM as the resources required are out of budget for all except a handful of companies in the world. Many other models do not need to be trained by you either as they are available already. There are several marketplaces where AI models can be obtained instantly on the internet (e.g., Hugging Face), often free of charge. The academic research on AI is mostly published online (e.g. in arXiv) before it is formally published in journals or conferences and is available free of charge. Many authors are willing to share data and models with you if you ask nicely.

In totality, therefore, many challenges that are more common in nature can be solved simply by assembling a set of external resources in the right way and without having to do any scientific study or innovation at all. Challenges that are specific to your niche may require some science and this is then likely to yield a competitive advantage to you. In those cases, it is probably best to keep that expertise in house.

Universities and Research Labs

"Relaxing is not a waste of time—it's an investment in well-being."
"The tradeoff between doing well and being well is a false choice."

ADAM GRANT, *HIDDEN POTENTIAL*

As a scientific discipline, AI evolves relatively slowly. Great new ideas and techniques are created once every few years. The transformer architecture that underlies the GPT craze of recent years is a good example of this, having been published in 2017, leading to public visibility more than five years later, and not having significantly evolved three years after that, despite nearly unprecedented attention. When people say that AI is evolving at a breakneck pace, they are referring to the commercial ecosystem of companies, products, and services on offer on the global market.

For AI leaders, it is important to mark the difference between scientific evolution of AI as a discipline and commercial evolution of AI as market. These two aspects must also be studied separately as they occur separately. This section will cover the scientific part and the next section will cover the commercial part.

The scientific evolution is mainly pursued by academic researchers at universities and large commercially funded research labs. That research is then mostly published in journals and conferences for public consumption. A significant number of these publications also publish their data and source code. Keeping up with the research can be quite difficult because of the complexity of the research and the sheer volume of publications.

The complexity of the research is evident in the jargon-based technical language of the published papers. They are often hard to read, understand, and put into a larger context. This requires you to have people on staff who are academically equipped to read these papers and who can quickly assess whether a particular paper is likely to be relevant to the company or not. Typically, I will read the abstract to determine what novelty this paper claims. If this is interesting, I will read the conclusion next to determine whether that novelty was achieved and to get an idea of its quality and range or application. Only if the result is applicable to current challenges and good enough to be practically relevant will I attempt to read the body of the paper. Most of the time, a cursory read is enough to determine that this is a dead end.

The volume of publications is a side effect of the well-known publish-or-perish principle of academia. This is a strange reality that everyone hates but everyone participates in and so preserves. Virtually all researchers I know are either tenured professors at research universities or want to be. Consequently, even if they are working for a commercial research lab or an enterprise that is looking at real-world applications of AI, they have all imbibed the academic spirit of publishing their research results. More so, they implicitly measure their worth as researchers by their publications. As it is difficult to compare the content of research papers between any two authors, the academic world adopted the number of papers as a proxy in the 1990s. Exceptional geniuses like Richard Feynman with 37 scientific papers over his entire career would never have obtained academic employment in this atmosphere.

Tracking towards a professorship is usually said to take approximately 4–6 academic papers per year. As most people do not work on research full-time, having other duties such as teaching and administration, each paper is worked on for approximately one full-time-equivalent month. Resulting from this, most papers do not represent a significant advance, or even a genuine difference to other papers—many papers get published several times under different titles.

All this being said, you need a strategy for dealing with this situation. On the one hand, your staff will want to publish and so you need a policy for that. On the other hand, you will want to learn the significant advances relevant to you and so will need a filter for the huge and complex volume of papers out there.

Two such filters will now be discussed in more detail.

University Research Relationships

There are multiple opportunities to establish mutually beneficial relationships with universities. Such relationships will provide you and your enterprise with access to people who are already spending a good part of their time filtering the literature and who have the expertise to do so. If you educate them about what you are looking for, they can become your prescreening filter over the large volume of publications out there.

These relationships also provide access to talent at all levels. Only some undergraduates become graduate students. Only some graduate students go on to do a PhD. Only very few PhDs become post doctoral students, and even fewer become tenure-track faculty. At each stage, the majority will leave academia to take a job on the outside. If you have maintained a good relationship with the university, they are likely to look at your company seriously.

One of the prime points about working with a university is to be aware of the university's timeline and policies. Working with students happens from September to May. Working with professors is often best in the summer months. PhD theses take three years, or more. Intellectual property is sometimes transferred to the student, sometimes to the university, and sometimes to the sponsor. Publication rights are often reserved. NDAs are sometimes a challenge and sometimes no problem. If your schedule and policies can adapt to their expectations, this could be the start of a beautiful friendship.

There are several ways to work with a university: Sponsor a master's or PhD thesis, conduct an undergraduate group project, subcontract custom research, or join a research consortium. Depending on your needs, they are all practical, relevant, and cost-effective ways to both get access to novel research and solve real problems. Let's look at each one of them in turn.

SPONSORING A THESIS

Sponsoring a thesis means different things at different places and often does not refer to spending any money at the undergraduate or master's level. Sponsoring a PhD student is typically on a fixed price schedule for a three-year commitment. What sponsoring means beyond money is mainly the setting of the challenge and regular contact to learn about progress. Often, it also means the provision of data, and perhaps it may mean the partial supervision of the student alongside the professor. The student and professor benefit from solving real problems, getting access to domain knowledge, and getting real-world data. The student also benefits from a chance of getting hired by you after the thesis work is done. You get a cost-effective attempt at solving a difficult problem.

CONDUCTING AN UNDERGRADUATE RESEARCH PROJECT

Keen to offer students a chance to work on real-life problems early on, many universities now offer a course over a full academic year in which a group of students—often five—work on a single project supervised by a professor and an industry sponsor. Having signed an NDA and providing the intellectual property to the industry sponsor at no cost, this is an excellent opportunity to outsource a research project if it is of a suitable size and scope.

SUBCONTRACTING CUSTOM RESEARCH

Research projects that are more advanced can be outsourced to professors directly in an expert consulting model. As opposed to the previous modes, the work here is done primarily by the professor with assistance from post-docs. Considering the time constraints due to other commitments, these projects ought to be difficult but not too time-consuming in sheer effort.

JOINING A CONSORTIUM

Research institutes at most research universities build consortia for certain large-scale challenges or topics. Consortia often last for many years and combine the research of several professors and dozens of PhD students and post-docs. In effect, they are mini departments in their own right. What sets them apart is a relatively narrow topical focus compared to a department but a wider lens than a single professorship or project. They are funded through industry membership, which is usually charged at a fixed yearly fee.

While the sponsors get access to the research and the staff, they typically do not directly set projects or supervise any work. It's an eyes-on but hands-off approach at a bargain price for a bundle of research. The value of this must be studied with care depending on your needs, particularly in view of the long timeline.

Apart from access to new research and a filter to the literature, the ways of cooperation are a great way to solve difficult problems in AI. Some challenges that you will face can be assembled from existing parts. Other challenges can be solved by your staff in a handful of months and so are accessible in an environment relevant for a commercial enterprise that is itself not an AI company. That leaves the question of what to do with challenges that cannot be solved in a short period of time by the staff you have. Most enterprises will not want to invest their limited resources into such projects. Outsourcing them to universities is the best realistic way to solve them. My personal rule of thumb is three months—if it is likely to take more than three months for a small group of my team to solve the problem, we declare it a research problem and either outsource it or deprioritize it.

Academic Conferences and Papers

Academic conferences are a great way to get compressed access to research. The presentations have been carefully selected by an expert panel and so you are seeing the cream of a typically large crop. As the presentations are live and in-person, you have immediate access to the researchers themselves if you have any further questions or want to get the data or code from them. Understanding the details is also often possible in far less time during a presentation than an attempt to read a dense academic paper. The extent to which the method is ready to be applied to a real-life situation can also be checked out quickly by directly addressing this with the researchers personally. In my experience, they will provide a candid assessment of their own results relative to solving a practical problem and are mostly open to help or even collaborate on making their techniques practically applicable.

The key is to select the right conferences to go to. There are a great many of them and you could easily spend all your time on conference tourism. With a little care, you can identify conferences that are essentially a publish-or-perish platform for PhD students to talk about their ongoing work or vanity-style conferences that draw no audience apart from the collective of speakers. What you are looking for are conferences professionally organized with a serious panel of experts looking over the proposed presentations after a call for papers has been put out, often six months or more ahead of the conference itself. Most of these events occur yearly, draw a large audience, and have an established track record of high quality. Many of them also have a more detailed scope than just AI and focus on either language models, computer vision, or some other branch of AI.

Before you go there, or send someone to go, it is a good idea to be clear on what you are looking for. You will be exposed to projects several times per hour and it is easy to get overwhelmed in that environment. Conferences are best treated like search problems. If you have search criteria in mind, you can easily assess each item for its fit.

Conferences often release what are known as proceedings. These are collections of papers written by the speakers at the conference. Not every presentation yields a paper, and a few papers will not have been presented due to author sickness and travel issues, so that there is not a strict one-to-one correspondence between the conference and its proceedings. Generally though, a proceedings is the written record of what was presented. Its main purpose is to put on a peer-reviewed record what the authors have done. The aspect of peer review is crucial for the academic career of the authors. To you, who are looking to solve problems, this aspect is largely irrelevant. The paper itself will probably have appeared on one of the online preprint archives weeks before and freely accessible to all.

Journals are the other accepted academic pathway to getting peer-reviewed work published. This is much less popular in AI research because it can take up to a year to be peer reviewed by journals as compared to a handful of weeks at conferences. The accepted way, in AI, is to submit your papers to prestigious conferences and get published in their proceedings. That is sufficient to have the paper credited in your academic career.

The actual dissemination of research happens via the preprint archives prior to the conferences and in-person at the conferences themselves. Of course, distribution of the links via social media is a popular way of marketing those primary channels.

So if you are interested in staying up to date on scientific work in AI, the best way to do this is to get regular updates from the preprint archives on certain search terms that are relevant to you, to go to a few carefully selected conferences, and to establish a trusted relationship with a research university.

Market Research Firms

"In a team setting, most people are hired for what they know and fired for how they apply what they know."

JOEL PETERSON, *ENTREPRENEURIAL LEADERSHIP*

Beyond academic research, AI presents us all with a unique challenge of market research. The speed at which new products or versions of products are released into the market and the volume of new companies offering new products is so great that a coordinated overview is practically impossible for an enterprise that does not have AI as its own focus.

It is clear that most offerings are either substandard, duplicates, or simply fabrications. Only a few offerings can withstand testing and are of lasting value to their customers. Identifying them quickly and cheaply is a significant problem.

Market research firms offer to do this work independently and objectively. They also investigate trends and best practices, so that you can learn from others' mistakes and avoid them. Such work is highly valuable, can save significant time, and reduces risks of missteps.

When faced with a particular challenge, you would want to know what companies out there have a product or service that can help. The pros and cons of these offerings are interesting. The track record and likely longevity of the company and the offerings are relevant. The extent to which the sales pitches are true and whether the vendors have done this before are important. All these aspects are covered by market research firms. Some of this content may be published by them in private access papers and some of it is accessible via consulting phone calls with analysts at the firm. In case you have special requests for content that does not yet exist, these firms can conduct custom market research for you.

In any case, membership with such a firm is likely to be better, faster, and cheaper than getting your own staff to do market research themselves.

In addition to one-way research that they provide, and you consume, many of these firms organize a community. Depending on the offer, this can range from an online chat group to a regular in-person meeting. These groups could be a handful of specially curated senior executives or a group of hundreds of like-minded leaders. Such communities are a fantastic way to learn from each other on what worked and what did not, which vendors provided good outcomes and which did not, and so on. Here is where you meet your peers in a private space where you can get trusted answers to practical questions. I highly recommend joining one or two such communities and regularly attending the meetings. As with all communities, the more you are prepared to proactively give it, the more it will provide back to you in return. Attending such events as a passive consumer will not provide full benefits.

Vendor Management

"The innovator's dilemma occurs because, when they first appear, innovations might not be good enough to serve the customers of the established companies in an industry, but they may be good enough to provide a new startup with enough customers in some niche to build a product. [...] Eventually, the startup has learned enough to create a strong product that takes away its larger rival's customers."

AJAY AGRAWAL, JOSHUA GANS, AVI GOLDFARB, *PREDICTION MACHINES*

At the enterprise level, you will need a variety of vendors for different aspects and so you need to manage a portfolio of vendors. The principal tasks are to admit a new vendor to the portfolio, to exit a vendor from it, and to choose a vendor from the portfolio for a particular item that you need. Apart from all the usual aspects of vendor management familiar from procurement processes that will not be treated in this book, there are a few unique aspects for AI vendors that will be addressed in this section.

Primarily, vendors will be differentiated into three categories for this book. Product vendors primarily sell access to a software product. Service vendors primarily sell work effort for some custom projects. Startups are a special category of significance for AI since there are so many startups, and they pose their own set of challenges that will be addressed.

Product Vendors

Companies that want to sell a product often make extravagant claims about the capabilities of that product. Demonstrations provided by the vendor look great but do not offer a picture of real performance because they are carefully prepared and choreographed to work well and to show the product in the most positive light. Such demonstrations, if not entirely useless, are only the very first entry point into a vendor assessment.

You should begin your process by clearly outlining your requirements, including hard requirements that you cannot live without and soft requirements for which you are prepared to live with partial fulfillment. If the data sheet or demo makes it clear that the product lacks a hard requirement, the judgment is very easy and you can exclude that vendor easily. That is the main benefit of a vendor-provided demo and is, usually, the full extent of what you can use the demo for.

To learn whether the product's features actually work well for you, you would have to test the product yourself. A full test in your environment would require full installation, which would usually require contracting and so a lengthy legal and technical review process for an enterprise. It is quickly clear that you cannot do this. You must find a way of testing without installation in your environment.

It is a best practice to generate a testing dataset at the same time as coming up with your requirements that you can use to test products. The purpose of this is that you want to have a set of data that you can give to vendors without needing to sign a non-disclosure agreement or any other kind of contract. Then you can test the product in the vendor's environment. While that is not a full test, it is so much faster and easier that it allows you to test more products than if you attempted to move towards a full internal installation.

To prevent yourself from getting side-tracked during testing at a vendor's offices or in the presence of the vendor's sales staff, it is good practice to have a written testing protocol with you. That way, you make sure that you cover all your points before you go home.

Obviously, you will want to talk to your colleagues from other enterprises about these vendors and get their feedback on any that they have tried out or worked with. Market research firms often have useful information also. Both of these sources combined often yield a realistic short list of companies for testing by your own staff.

Beyond the product itself, you must look at two other main artefacts: The roadmap and the service offering.

The **roadmap** is a depiction of future features or enhancements that the vendor plans to make. If an important feature that you need is coming in the next release, which is three months away, perhaps you are prepared to give them the benefit of the doubt if everything else is great today. If that important feature is not on the roadmap, you will want to have a conversation as to the amount of influence you will have over the roadmap if you become a client.

Especially for small vendors or large enterprises, the amount of roadmap influence could be significant. This conversation is important as it shows how important you are to them as clients, relative to the other clients. If you would be the most influential client, this sounds good but in practice also means that most of the bug finding will be done by you. If you are one of many and have no real influence, then you know you will have to take what you are given. Ideally, you are in the top 20 clients so that you get a seat on the customer advisory board but are not the single largest client. If this seems within reach, ask for a seat on the customer advisory board as this gives you direct influence over the roadmap and deeper insight into the vendor's plans.

A good product roadmap does not only have future feature descriptions on it but also explicit timelines when they will be available. Conversations with the technical staff of the vendor can be used to dive more deeply into what the vendor means by certain features or why some elements you might expect are not there. Sometimes, it turns out that what you want is either included in another item or has simply been given a different name.

The **service offering** by a product vendor is often overlooked by both vendor and client but can be as important as the product itself. The ability of the vendor to install and maintain the product is important. The willingness of the vendor to train your staff in the best usage of the product can be essential in unlocking the benefits of the product in the first place.

Standard software written for many clients is unlikely to provide you with precisely what you want. So, you will either have to live with it or customize it to your needs. Often, the only company capable of customizing the product is its vendor. So, the ability and affordability of this offering can be crucial in filling any gaps, no matter how small they might be.

The way the vendor treats you and your needs can be a significant enabler. I have personally and knowingly bought inferior products because the service was significantly better and the vendor took my concerns more seriously than their competitors did. In the end, you are buying a product not to have a fancy tool but rather to solve a—often long-term—business problem. That must be your primary focus, and service can be a large part of that solution.

Speaking of solutions, there are vendors that attempt to sell a full solution. This can be a double-edged sword. For example, there are companies that sell a computer vision solution that consists of the trained AI models and the cloud-enabled software that executes those models but the solution also contains the cameras, the local data acquisition, the networking that may be satellite based, and everything in between to move from rugged reality all the way to the outcome.

If you are looking for this solution and nothing else, this is a dream come true because the vendor has solved all the procurement, integration, and installation challenges for you. You can buy the no-worries package.

Most enterprises, however, due to their size and complexity, will have several similar problems to solve. Buying a hardware bundle for problem one and another hardware bundle for problem two is not ideal. If you have a dozen problems of this kind, the expense and additional complexity of maintaining all the hardware can be prohibitive. Here, you may want to separate the purchasing of hardware, software, and models and choose to do the integration as a custom project.

Computer vision is the primary candidate for this because of the expense of cameras and the bandwidth requirements for streaming video or just sending still images at regular intervals. But industrial sensors can be costly and difficult to install also, especially applications involving vibration signals. Networking in areas that are not covered by regular internet coverage can create significant challenges that you may need to also solve independently.

While choosing a vendor can be complex, it is good to maintain a relationship with the vendor so that you stay informed and provide your insights back. The vendor will gain from your feedback, and you will gain from their information and the resulting special treatment. If the product is at all important to you, it is of great benefit to treat the vendor like a partner in solving your business problem rather than a transactional point of sale for a widget. Vendors are often surprisingly helpful in getting things done if only they can be part of the journey and solution.

Service Vendors

Services companies are different in most ways from product companies in that they, in various forms, supply a temporary workforce. As they supply people and often draw on the same labor pool, it can be hard to distinguish between companies in abstraction of the individuals that the service company will assign to your project. Some clients are therefore tempted to treat this like recruitment and so look at resumes and conduct interviews.

I do not recommend this recruitment approach but rather recommend a long-term relationship with their account team and a more senior practice or team leader. They will then have the responsibility to select the best people from their pool for your project. If they have a relationship with you that is not transactional but long-term or even strategic, they will do a better job of selection that you can realistically do from the outside.

If the scope is sufficiently large, the important thing is not which mistakes the team ends up making but rather again the trusted relationship so that you can rely on the team and their leader to spot and fix those mistakes over the course of the project.

Somewhat jokingly, I often ask what comes first, the end of the budget or the end of the project. If at the end of the project there is some budget remaining, all is well and the good relationship with the services vendor can continue to the next project. If at the end of the budget there is some part of the project remaining, then we often have difficulties. Even if you want to increase the budget, you may not be able to due to enterprise budgeting constraints and so that situation may amount to a total failure, even if the gap is small. This basic conversation must be had, and the vendor must understand the boundary conditions that apply to them.

There are two kinds of services contracts. In the first, the vendor owes you a certain number of labor hours, irrespective of the result. Guidance towards the goal you have in mind is your problem. In the second, the vendor owes you a concrete outcome, irrespective of the number of hours or the path taken. Explicit definition of what the outcome must be is your problem here and you may end up fighting about items that the vendor sees as scope creep. Generally, the second costs more money because the vendor accepts the risk of failure, but you have additional leverage in case something goes wrong or things take longer due to unforeseen circumstances.

Here, too, it is good practice to ascertain how seriously the vendor takes a relationship with you. If they are also invested in making this work long term and so doing a string of projects together, you have a much higher chance of getting a good outcome than not. From this perspective, ending up with success does not depend that strongly on the individuals on the team but more on the relationship with the company and its leadership and account team as they can rearrange things internally to make sure you are happy.

The practical consequence is that you cannot maintain many such relationships and will have to say no to most vendors on virtually no grounds other than the fact that they do not have a close relationship with you. This sounds much like the chicken-and-egg situation and is not ideal, but realistic.

Startups

As mentioned in Chapter 2, startups are a special kind of company whether they are in the product or service category. Apart from the considerations mentioned in Chapter 2 or above in this section, startups are harder to assess because they lack a track record. They are also harder to compare because their economic model can be quite different from those of their competitors—for example, they might be prepared to lose money on you just to have you as a client.

The major problems are that their offering is not stable and is subject to change at a moment's notice as the founders pivot several times in their attempt to find the optimal market position for themselves. Sometimes this is driven by you, which is great. Sometimes you can be left behind in a pivot driven by other forces, which can effectively eliminate the product that solved your problem, and you may now have a bigger problem than before.

If you are a sufficiently large client of the startup or a sufficiently large enterprise, you must expect at some point to be asked if you want to buy either a significant stake in the startup or the entirety of the startup. It is expedient to form an opinion on basic questions like this in advance because when the question is asked, an answer must usually be forthcoming in days or a handful of weeks before the opportunity is lost.

For startups, more so than for other companies, a relationship with the people—principally the founding team—is important. Recall that a company is a startup primarily because it seeks growth above everything else. If the leaders of the startup see their growth in the direction of solving your problems, then any direction you can provide will be good for both of you. Whether or not they see this can depend significantly on you whether you have a relationship with them. In this way, the inherent nature of startup pivoting can be leveraged in your favor.

Open Innovation

"The process of escalating commitment: The more time, energy, and resources that get invested in the negotiation and due diligence process— the harder it is to walk away."

JEFFREY PFEFFER, WHAT WERE THEY THINKING?

Open innovation is the approach of solving the enterprise's challenges using a number of external resources such as individuals, research institutions, and vendors of different kinds. The enterprise can benefit from innovation that happens with those partners, or it can co-create innovation with them. It is often a good idea to contribute back to that public community in the form of publications in writing and presentations at events. The innovation is therefore not closed—happening strictly in-house—but open and happening through a network of different partners.

As mentioned, it is good practice to focus on the long-term relationship between the enterprise and those partners as opposed to a transactional process. In virtually every case, the quality of the outcome is significantly better if the people on the other side feel comfortable and treated with respect and mutual understanding.

Another interesting aspect is cooperation with your own competitors. There are always areas that are not competitive in themselves, and these could be interesting to all if data and effort were shared between companies that are, on the face of it, competitors. A simple example for many producing industries is the safety of their workers, which is important for all equally. Perhaps a consortium could be formed that exchanges data to make an industry model, or that exchanges models.

The modern technique of federated learning can be helpful here because no partner gets to see the data of any other partner directly, but all partners benefit from the model that has been trained on everyone's data. This is an approach gaining traction, particularly in the healthcare industry where personally identifiable data makes it difficult to share data outright.

Technology is made, used, and bought by people. Human relationships of trust matter. The ability to cooperate and work as a team, even between companies, makes for better outcomes than anyone can achieve by themselves. That is **open innovation**. Especially for AI, which is driven principally by access to data, computational resources, and bright ideas, collaboration is key.

KEY TAKEAWAYS

1   Decide on some high-level ground rules on how your enterprise will navigate the buy-versus-build debate in relation to AI. These will serve you well as relevant decisions will need to be made often.

2   Develop relationships with academic institutions and curate a list of relevant academic and industry conferences which the AI team will attend. Included in this is a practical plan for how the team will keep up with the stream of research output in this vibrant field.

3   Set up a function to manage relationships with vendors with a view to achieving a trusted partnership with people who understand your enterprise, its industry, and its business for faster and better outcomes.

15

Lifelong Learning for the Team and Company

AI is a topic that almost everyone has an opinion about and has heard a lot about but does not understand. Somewhere between Hollywood movies and endless sales pitches by cloud providers and startups, the general public is lost. Most stories told about AI quickly yield either a utopian or dystopian view of the near-term future and thus generate emotions from fear or excitement to greed. The board members of your enterprise are likely to ask their grandchildren for help with AI when push comes to shove.

When compared to many other topics in software, IT, or general technology, AI is fairly unique in sheer widespread presence and level of general interest. That is not necessarily a good thing. The narrative is out of control.

It is a core responsibility of a central AI program at an enterprise to regain some semblance of control over the AI narrative, to put the risks and opportunities into context, and to steer the conversation to realistic attainable goals. You will have three main audiences for any educational efforts inside the enterprise: all employees, leaders, and citizen data scientists and AI experts.

Education for Everyone at the Enterprise

"There are three ways leaders typically fall short in their communication efforts. The first is quantity. Most simply don't communicate enough."
JOEL PETERSON, *ENTREPRENEURIAL LEADERSHIP*

As AI is a cross-functional discipline that, in some form, may touch every employee at the enterprise, it is a good idea to make training available to everyone. This training would focus more on the scientific concepts and usage risks of commonly available tools. Understanding the economic and managerial aspects of AI discussed in this book is less important here.

The most common general misconceptions that I often find are:

1 AI will take away my job.

2 AI will take over the world—like in Hollywood movies such as *The Terminator*.

3 AI is intelligent and will soon achieve superintelligence.

The AI training will do well to emphasize that while AI may be able to automate individual tasks, any one employee performs many tasks as part of their job. Repetitive and isolated tasks that occur often can be automated more easily than rare and integrated tasks. It is generally virtually impossible to replace a holistic job with AI due to the diversity of tasks that an employee usually performs. In my view, the idea that AI will automate away a significant portion of jobs is simply false for the foreseeable future. The only exceptions may be jobs that are inherently very narrow in their scope, such as driving a car or truck. Even here, it has turned out to be much harder to replace these individuals than the AI industry initially claimed. One might debate this at great length, but the conclusion that I want to emphasize is: Do not worry!

For AI to take over the world, it would not just need to be intelligent—I will get to that— it would also have to be able to physically interact with the world, just like the Terminator was able to do in the movie. However advanced you believe AI to be, robotics is far behind, and the general connectivity of the world is even further behind. Much of the world's essential infrastructure and most of the military capability cannot be hacked because it is not connected in the way it would need to be. Most of the scenarios popularized in blockbuster movies in the last ten years would not be possible, even if there were a superintelligent and malicious AI entity.

AI is not intelligent. What AI does is learn a pattern by means of looking at a great many examples of that pattern, and possibly other distinguishing patterns. It learns to recognize a cat by looking at pictures of cats and pictures that are not of cats. It learns to construct sentences by looking at many sentences and studying the probabilistic patterns of how words follow each other based on the context that came before. There are many definitions of intelligence—discussing that is beyond the scope of this book—and none of them applies to the AI systems in existence today. In the words of theoretical physicist Stephen Hawking, "Intelligence is the ability to adapt to change."[1] In that sense, AI is quite dull.

Large language models seem to some to exhibit signs of intelligence. That is coincidence because for the first time, AI can communicate with us in the way we communicate with each other. The way humans create speech is fundamentally different from AI, however. AI strings words together using probabilistic patterns. Humans string words together based on causal, logical, emotional, and intentional reasoning and generally elicit some effect. The effect may be simple like a laugh, or complex like an agreement to purchase a software license.

The fact that the current LLMs are not there yet is pretty evident. How far away are we? There are three main ways to improve AI models in general, and LLMs specifically. First, get more data. LLMs have been trained on more or less all the data available already and so that avenue is closed. Second, get more computational resources. The current models already cost several hundred million dollars in computational resources per training run. While one could spend more, the ceiling is very nearly reached. Third, scientific innovation. To AI specialists it is clear that the transformer architecture, while clearly revolutionary, is not capable of the kind of intelligence AI has been dreaming of. New architecture and new algorithms are required. That is the avenue that is open and the avenue that is being pursued by the few companies that have the financial resources to do so. However, revolutionary innovation cannot be scheduled and so no timeline can be stated. It is also likely that a single innovation will not be enough but that multiple steps building on each other are going to be needed to get there, just like the transformer, by itself, did not bring about the GPT revolution of today.

That brings me to my rough conclusion: The AI community is five Nobel prizes away from intelligence—if it will ever get there. In terms of time, that will most likely extend past my lifetime.

AI Education for Leaders and Executives

"C-suite leadership is required for AI strategy [because] ... the implementation of AI tools in one part of the business may also affect other parts."

AJAY AGRAWAL, JOSHUA GANS, AVI GOLDFARB, *PREDICTION MACHINES*

Leaders and executives in the enterprise need to understand some rudimentary concepts of AI itself but mainly need to imbibe the concepts discussed in this book. They need to deeply understand the ecosystem of actions and dependencies in which AI is made and used. The major misconceptions that leaders tend to have are as follows:

1 AI is automatic and therefore ready to go in days, with virtually zero costs to create.

2 AI can be "plugged into" the processes and IT applications of the enterprise with little effort.

3 Having been plugged in, AI just works and people then go about their day more efficiently.

While the opinions may be substantially more nuanced than this, they commonly boil down to these after a little debate. These are eminently reasonable and understandable opinions because that is what most of the vendors constantly say in their sales pitches. Alas, they are not true, as you now know.

The fact that the development of models and software takes time and resources is something that is usually not hard to explain. The scientific and mathematical concepts of AI are also not difficult and do not present a challenge. Change management and the need to focus deeply on user experience is harder to grasp. Software-related change management is much harder than other change management because software usually has an order of magnitude more moving parts than, for example, introducing a new physical machine into a factory. Additionally, the number of people involved in software changes is often far greater.

Dependency on data is an eye-opening experience for many leaders. Used to getting and analyzing data through human domain experts who know all the workarounds needed to get their executives what they need, it is hard to discover that a prerequisite for AI may be a month-long data-transformation exercise.

The point at which it becomes challenging is ethics, risk, and governance. Due to the cross-functional nature of AI, the drive for governance can seem to some to be a power grab by the AI team. Getting alignment on an enterprise-wide AI governance process that reviews and approves use cases and applications is probably the hardest challenge of setting up an enterprise AI program. The ability to do this is predicated on educating the stakeholders in all the ramifications discussed above.

Apart from initial training on all these concepts, it is important for leaders to be kept in the loop on what the AI program is doing. In particular, the use cases deployed, and their business value achievements, are crucial to communicate. To this end, the AI program may choose to release a regular report to the leadership covering the bottom-line achievements. While the AI program may not formally be a business unit with an internal profit-and-loss statement, it should run itself in that spirit nonetheless and make that attitude clear in its regular reporting.

Education for Citizen Data Scientists and AI Experts

"Automation that eliminates a human from a task does not necessarily eliminate them from a job."
        AJAY AGRAWAL, JOSHUA GANS, AVI GOLDFARB, PREDICTION MACHINES

The group of citizen data scientists and AI experts are the people who will actually make AI happen in the enterprise. In order to belong to that group, they will already have had some form of relevant education and training. Yet, it is important to continue that training. For this group, there are three primary topics of education:

1 The conceptual and mathematical foundations of AI.

2 Software engineering and computer programming languages like Python and code libraries like PyTorch.

3 The use of a plethora of diverse software tools and cloud-based programs to make, deploy, assess, and monitor the models and applications.

There are more books and courses on the first two groups that anyone could read or watch in a lifetime and so those are well taken care of. The main duty of the AI team is therefore to make a good selection and distribute them or directly teach those concepts and skills to the general audience at the enterprise. The aim is mainly to bring a diverse population up to some well-defined common standard of skills. Relative to computer programming, an advanced AI program will have coding standards that will be taught and adhered to as well.

The third main topic is usually well taken care of by the appropriate vendors of these tools, most of whom also supply certificates for graduating from their training programs.

In addition to the technical skills, it is important for this group to understand and follow the processes of the enterprise AI program in development and deployment so that the final vision is achieved—real users using an AI application and generating real-world benefits for the enterprise.

The best way to achieve all of this is to provide regular sessions at which projects and models are presented and discussed from real enterprise use cases made by a member of this group. That allows the group to teach itself while simultaneously seeing the techniques in action for a use case that is deployed at the company.

AI News Service

"If you … don't know the answers to key questions, don't fake it."

JOEL PETERSON, *ENTREPRENEURIAL LEADERSHIP*

The AI world evolves quickly. While AI science evolves quite slowly, the commercial ecosystem of companies and software versions develops at a pace that is virtually impossible to keep up with. Having an enterprise internal news service for AI-related news and events is very helpful. The most important element here is to set any news into context. Two contexts are particularly relevant. On the one hand, the news should be de-hyped to bring it down to reality and to examine to what extent any claims are true. On the other hand, how the news is relevant to the enterprise should be outlined. Most AI news is not relevant to the enterprise. That context setting alone will prevent much time being wasted.

The news service would also highlight any enterprise AI news, such as successes of internal use cases, applications developed and released, projects in flight, and people who are doing great things with AI.

Finally, encouraging a dialogue of questions and answers is helpful so that people around the company feel they know where to go if they have AI questions and will get a trustworthy answer.

Community Events

"The confidence that individuals have in their beliefs depends mostly on the quality of the story they can tell about what they see, even if they see little."

DANIEL KAHNEMAN, *THINKING, FAST AND SLOW*

AI as a topic attracts interest from most people at the enterprise in some form. Your enterprise AI program will have generally interested people, users of the applications you release, domain experts and leaders working with you on projects, and a sizable population of citizen data scientists directly working with the AI team on models and applications. These communities want to hear from you. Releasing regular reports discussing what is going on will be well received.

When something good has occurred or someone has done something great, celebrate the victory, present awards, and tell the community. This is a learning moment as well as a morale-building and interest-generating moment that allows the whole program to gather forward momentum. I am consistently amazed how much traction one can achieve by those magical phrases like "Thank you," "I appreciate you," or "You have done a fantastic job."

In-person events where they can hear from the AI team and their fellows are excellent to foster a common view on this modern technology and its place in the enterprise. These events can be training sessions, regular project presentations, topical discussions on particular aspects of AI, or internal podcasts on projects.

A particularly great way is to hold a yearly internal AI conference where the community gets together and systematically reviews what the enterprise is doing. People hear about different projects in many different parts of the company. They broaden their horizons, not just about AI but about the enterprise itself. A crucial part of AI is the realization that it is "just" an enabling technology for value in all parts of the company.

An enterprise AI program is about the enterprise, not about AI. Whenever faced with a decision, I tell myself, "It is not about me" and ask myself, "What action gets us closer to value?"

KEY TAKEAWAYS

1 Create training courses for everyone at the company to upskill them in AI, to calm their fears, and to set the context of what the enterprise wants to achieve with it.

2 Host training courses for leaders and executives to raise awareness of what is needed to make and deploy AI as well as to keep it running. Pushing past the hype, they need to understand the real costs and benefits of AI in the context of the enterprise's current status.

3 Keep the entire enterprise up to date on the relevant AI news, both in the world and at the enterprise. Celebrate your wins.

Note

1 Black Holes Talk, Oxford University, 2017.

FURTHER READING

The following books are not only the sources for some of the quotes in this book but are resources from which I have learned a great deal on the general topic of management. I most heartily recommend reading each and every one of these, with a large cup of tea, on any cold and rainy Sundays that you may have coming up.

Jeffrey Pfeffer, *What Were They Thinking?*, Harvard Business School Press, 2007.

Brené Brown, *Daring Greatly*, Gotham Books, 2012.

Seth Godin, *Tribes: We need you to lead us*, Portfolio, 2008.

Arnold Schwarzenegger, *Be Useful: Seven tools for life*, Penguin Press, 2023.

Carol Dweck, *Mindset: The new psychology of success*, Random House, 2016.

Stephen M.R. Covey, *The Speed of Trust*, Free Press, 2006.

Simon Sinek, *Start With Why*, Portfolio Penguin, 2009.

Ajay Agrawal, Joshua Gans, and Avi Goldfarb, *Power and Prediction*, Harvard University Review Press, 2022.

Ajay Agrawal, Joshua Gans, and Avi Goldfarb, *Prediction Machines*, Harvard University Review Press, 2018.

Marshall Goldsmith, *What Got You Here Won't Get You There*, Hachette Books, 2007.

Vin Vashishta, *From Data to Profit*, Wiley, 2023.

Paul J. Zak, *Trust Factor*, Amacom, 2017.

Adam Grant, *Hidden Potential*, Viking, 2023.

Michael Marquardt, *Leading With Questions*, Jossey-Bass, 2005.

Bill Schmarzo, *The Economics of Data, Analytics, and Digital Transformation*, Packt, 2020.

Rahaf Harfoush, *Hustle & Float*, Diversion Books, 2019.

Marvin Weisbord and Sandra Janoff, *Don't Just Do Something, Stand There!*, Berrett-Koehler, 2007.

Paul Leinwand, Cesare Mainardi, and Art Kleiner, *Strategy That Works*, Harvard Business Review Press, 2016.

Joel Peterson, *Entrepreneurial Leadership*, HarperCollins Leadership, 2020.

Jeffrey Pfeffer, *7 Rules of Power*, Matt Holt Books, 2022.

Daniel Kahneman, *Thinking, Fast and Slow*, Farrar, Straus and Giroux, 2013.

Olivia Gambelin, *Responsible AI*, Kogan Page, 2024.

The following resources are also sources for some quotes in this book and have excellent content, but their wider topics are somewhat more tangential to the main focus here than the resources listed above.

Raffaella Sadun, Nicholas Bloom, and John van Reenen, Why do you undervalue competent management?, *Harvard Business Review*, September 2017.

Steve Blank, What's a startup? First principles. Blog post from January 25, 2010. https://steveblank.com/2010/01/25/whats-a-startup-first-principles (archived at https://perma.cc/UPL5-HHZL)

Eric Ries, *The Lean Startup: How constant innovation creates radically successful businesses*, Viking, 2011.

Paul Graham, Startup = Growth. Blog post from September 2012. https://www.paulgraham.com/growth.html (archived at https://perma.cc/MN4J-VRUE)

Warren Bennis and Burt Nanus, *Leaders: Strategies for taking charge*, Harper & Row, 1985.

Richard Hamming, *The Art of Doing Science and Engineering: Learning to learn*, Stripe Press, 2020.

Thomas S. Kuhn, *The Structure of Scientific Revolutions*, Second Edition, University of Chicago Press, 1970.

Matthew Dixon and Brent Adamson, *The Challenger Sale*, Portfolio Penguin, 2011.

Nathan Rosenberg, *Inside the Black Box: Technology and economics*, Cambridge University Press, 1982.

Michael Shermer, *The Believing Brain*, Times Books, 2011.

Tom Mitchell, *Machine Learning*, McGraw-Hill, 2013.

Harold Abelson, *Structure and Interpretation of Computer Programs*, The MIT Press, 1996.

INDEX

Note: Page numbers in *italics* refer to figures or tables.

Looking for another book?

Explore our award-winning books from global business experts in Digital and Technology

Scan the code to browse

www.koganpage.com/digital-technology

More from Kogan Page

ISBN: 9781398615700

ISBN: 9781398619555

ISBN: 9781398618008

ISBN: 9781398622012

From 4 December 2025 the EU Responsible Person (GPSR) is:
eucomply oÜ, Pärnu mnt. 139b – 14, 11317 Tallinn, Estonia
www.eucompliancepartner.com

www.ingramcontent.com/pod-product-compliance
Lightning Source LLC
Chambersburg PA
CBHW070940050326
40689CB00014B/3279